MEMOIRS
of the
American Mathematical Society

Number 544

W0017446

The 2-Dimensional Attractor of
$$x'(t) = -\mu x(t) + f(x(t-1))$$

Hans-Otto Walther

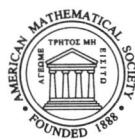

January 1995 • Volume 113 • Number 544 (end of volume) • ISSN 0065-9266

American Mathematical Society
Providence, Rhode Island

1991 *Mathematics Subject Classification.*
Primary 34K15; Secondary 58F12.

Library of Congress Cataloging-in-Publication Data

Walther, Hans-Otto.
  The 2-dimensional attractor of $x'(t) = -[mu]x(t) + f(x(t-1))$ / Hans-Otto Walther.
    p. cm. – (Memoirs of the American Mathematical Society, ISSN 0065-9266; no. 544)
  "Volume 113, number 544 (end of volume)."
  Includes bibliographical references.
  ISBN 0-8218-2602-6 (pbk. : alk. paper)
    1. Delay differential equations. 2. Differentiable dynamical systems. I. Title. II. Title: Two-
dimensional attractor of $x'(t) = -[mu]x(t) + f(x(t-1))$. III. Series.
QA3.A57   no. 544
[QA371]
510 s–dc20
[515\.352]   94-36555
                                                                                          CIP

Memoirs of the American Mathematical Society

This journal is devoted entirely to research in pure and applied mathematics.

Subscription information. The 1995 subscription begins with Number 541 and consists of six mailings, each containing one or more numbers. Subscription prices for 1995 are $369 list, $295 institutional member. A late charge of 10% of the subscription price will be imposed on orders received from nonmembers after January 1 of the subscription year. Subscribers outside the United States and India must pay a postage surcharge of $25; subscribers in India must pay a postage surcharge of $43. Expedited delivery to destinations in North America $30; elsewhere $92. Each number may be ordered separately; *please specify number* when ordering an individual number. For prices and titles of recently released numbers, see the New Publications sections of the *Notices of the American Mathematical Society*.

Back number information. For back issues see the *AMS Catalog of Publications*.

Subscriptions and orders should be addressed to the American Mathematical Society, P. O. Box 5904, Boston, MA 02206-5904. *All orders must be accompanied by payment.* Other correspondence should be addressed to Box 6248, Providence, RI 02940-6248.

Memoirs of the American Mathematical Society is published bimonthly (each volume consisting usually of more than one number) by the American Mathematical Society at 201 Charles Street, Providence, RI 02904-2213. Second-class postage paid at Providence, Rhode Island. Postmaster: Send address changes to Memoirs, American Mathematical Society, P. O. Box 6248, Providence, RI 02940-6248.

Contents

Abstract

Let a smooth real function f be given which satisfies

$$\xi f(\xi) < 0 \quad \text{for} \quad \xi \neq 0$$

and is bounded from above or from below. Let $\mu \geq 0$. The equation

$$x'(t) = -\mu x(t) + f(x(t-1))$$

generates a semiflow F on the phase space $C([-1,0], \mathbb{R})$, which leaves the set S of data $\phi \neq 0$ with at most one change of sign invariant. (The domain of absorption into S is open, and it has been proved to be dense in special cases.) The induced semiflow on the complete metric space $\overline{S} = S \cup \{0\}$ has a global attractor A, which is a subset of the global attractor of the full semiflow F. If in addition $f'(\xi) < 0$ for all ξ, and if $A \neq \{0\}$, then A is a 2–dimensional Lipschitz continuous graph which is homeomorphic to the closed unit disk; the unit circle corresponds to a periodic orbit in A. In the course of the proof we derive a Poincaré–Bendixson theorem and show that periodic orbits in A are nested into each other.

Key words and phrases. Differential delay equation, attractor, slowly oscillating solutions, planar dynamics.

CHAPTER 1

Introduction

Let a continuous function $f : \mathbb{R} \to \mathbb{R}$ be given which satisfies

$$\xi f(\xi) < 0 \quad \text{for} \quad \xi \neq 0,$$

and let $\mu \geq 0$. The equation

$$(1.1) \qquad\qquad x'(t) = -\mu x(t) + f(x(t-1))$$

is the simplest differential equation for a state variable under the influence of delayed negative feedback with respect to an equilibrium ($\xi = 0$), and friction. We shall assume in the following that f is continuously differentiable, with $f(0) = 0$ and $f'(\xi) < 0$ for all ξ, and $\sup f < \infty$. These conditions are satisfied for a number of equations that arise in applications. An example is Wright's equation

$$x'(t) = -\alpha x(t-1)[1 + x(t)], \quad \alpha > 0$$

whose solutions with range in $(-1, \infty)$ can be transformed into the solutions of the equation

$$x'(t) = \alpha(1 - e^{x(t-1)}).$$

Let C denote the space of continuous real functions on the interval $[-1, 0]$, equipped with the supremum–norm. Each $\phi \in C$ uniquely determines a function $x^\phi : [-1, \infty) \to \mathbb{R}$ which coincides with ϕ on $[-1, 0]$ and satisfies eq. (1) for all $t > 0$. The relations

$$F(t, \phi) = x_t^\phi \quad \text{for} \quad t \geq 0, x_t^\phi(s) = x^\phi(t+s) \quad \text{for} \quad -1 \leq s \leq 0$$

define a semiflow F on C.

We are interested in the asymptotic behaviour of solutions which are slowly oscillating in the sense that consecutive zeros are spaced at distances greater than the delay 1. This is motivated by the fact that the set C_s of initial data ϕ so that the solution x^ϕ is slowly oscillating on some unbounded interval is open, and by the conjecture that C_s is dense [9]. For a proof of the latter in special cases, see [18]. The object associated with slowly oscillating solutions in the phase space C is the set S of data $\phi \in C \setminus \{0\}$ with at most one change of sign. All segments x_t of slowly oscillating solutions belong to S. S and its closure

---

Received by the editor September 2, 1992.

$\overline{S} = S \cup \{0\}$ are positively invariant under the semiflow, so that F induces a semiflow on the complete metric space $\overline{S}$. We shall see in Chapter 4 that the induced semiflow possesses a global attractor $A \subset \overline{S}$. The attractor A contains the stationary state $\phi = 0$ and, for example, all orbits in C of slowly oscillating periodic solutions.

Our main result, Theorem 7.1, says that in case A is nontrivial, i.e. $A \neq \{0\}$, it is a Lipschitz continuous graph of dimension 2, bordered by a periodic orbit and homeomorphic to a closed disk in the plane. The semiflow on A extends to a complete flow.

In light of the density conjecture mentioned before it may be said that the planar dynamics on A govern the typical long–term behaviour of solutions of eq. (1.1).

Sufficient conditions for A to be nontrivial are for example that the stationary state is linearly unstable (compare Chapter 5), or that there exists a slowly oscillating periodic solution (see [8] and the references in [19]). The latter is in fact equivalent to A being nontrivial, by Theorem 7.1.

The full semiflow F on C has a global attractor $A(F)$. The relation of $A(F)$ to A is as follows. There are examples where A is nontrivial and

$$A = A(F),$$

see e.g. [21]. In other cases, however, A is a proper subset of $A(F)$, and

$$\dim A(F) > 2 = \dim A$$

since $A(F)$ contains the unstable set of the stationary state whose topological dimension is a large even number for suitable μ and $f'(0)$.

The attractor $A(F)$ is easily characterized as the set of segments x_t of bounded solutions which are defined on $\mathbb{R}$. For the collection of these solutions, J. Mallet–Paret found a Morse decomposition [11]. If the stationary state of F is linearly unstable then the most stable invariant set of this Morse decomposition is the set S_1 of those bounded slowly oscillating solutions on $\mathbb{R}$ which satisfy

$$\liminf_{|t| \to \infty} |x(t)| > 0$$

The relation of A to S_1 is that A consists of all segments of solutions in S_1, together with $\phi = 0$ and all heteroclinic phase curves in the set S which connect 0 and a phase curve of a periodic solution in S_1.

The second main result of the present paper, Theorem 10.1, is of the Poincaré–Bendixson type. It says that α– and ω–limit sets of phase curves in A different from the singleton $\{0\}$ are periodic orbits, given by slowly oscillating periodic solutions. Related results have been, or will be, obtained also in other work [14, 13, 12, 16], by methods which are different from the one developped here.

Organization, basic concepts, tools. The preliminary Chapter 2 provides, among others, elementary facts on angles along curves in the plane, which will be used in the proofs of Theorem 7.1 and Theorem 10.1.

Chapter 3 deals with the semiflow F and slowly oscillating solutions. The strict monotonicity of f implies that all maps $F(t, \cdot)$ are injective, and that the set S is also positively invariant under *differences of phase curves*. On a cone $K \subset S$, the semiflow defines a global return map P; phase curves of bounded slowly oscillating solutions $x : \mathbb{R} \to \mathbb{R}$ intersect transversally with a hyperplane $H \supset K$.

In Chapter 4 the attractors $A(F)$, $A \subset \overline{S}$, and the attractor $A(P)$ of the map P are introduced and characterized. We find

$$A = A(P) \cap \overline{K}.$$

Chapter 5 recalls facts about the linearization of F at the stationary point 0. The semigroup of the maps

$$T(t) = D_2 F(t, 0), \quad t \geq 0$$

or, the linearized equation

$$(1.2) \qquad x'(t) = -\mu x(t) - \alpha x(t-1), \quad \text{with} \quad \alpha = -f'(0),$$

define a decomposition

$$(1.3) \qquad\qquad C = L \oplus Q$$

into $T(t)$–invariant subspaces which will later be used for the representation of A as a graph. The space L in (1.3) is the 2–dimensional reellified eigenspace of the leading pair of points in the spectrum of the generator of the semigroup, and Q is the reellified eigenspace associated with the rest of the spectrum. We have

$$L \setminus \{0\} \subset S$$

since $L \setminus \{0\}$ consists of segments of slowly oscillating solutions of the linear equation (1.2), and

$$Q \cap S = \emptyset.$$

In terms of the projection p onto L given by (1.3), the last equation is equivalent to

$$(1.4) \qquad\qquad 0 \notin pS.$$

This relation reduces the proof that the attractor A is given by a map $a : pA \to Q$ to the proof that

$$(1.5) \qquad \phi - \psi \in S \quad \text{for all} \quad \phi, \psi \quad \text{in} \quad A \quad \text{with} \quad \phi \neq \psi$$

— if (1.5) holds, and if ϕ, ψ are different points in A, then $\phi - \psi \in S$, and by (1.4),

$$0 \neq p(\phi - \psi) = p\phi - p\psi;$$

thereby the restriction $p|A$ is injective, and the assertion about A follows.

Chapter 6 begins with an a–priori estimate of segments along differences of slowly oscillating solutions, taken from [**19, 20**]; in its simplest form it goes back to [17]. The a–priori estimate will later imply that the graph A is Lipschitz continuous. The remainder of Chapter 6 prepares the derivation of (1.5) in a peculiar situation, namely when ϕ and ψ sit on phase curves which both tend to 0 as $t \to -\infty$. We prove that such phase curves belong to an invariant manifold which is tangent to L at $\phi = 0$. The manifold in question is either a center manifold, or the manifold W_0 discussed in [19], depending on the location of the leading pair in the complex plane.

Chapter 7 contains Theorem 7.1, the main result, and the proof that A is a 2–dimensional Lipschitz continuous graph with respect to the decomposition (1.3).

The proof of the remaining assertion that in case $A \neq \{0\}$, pA consists of a projected periodic orbit and its interior requires further preparations. In Chapter 8 we use the flow on A to construct homeomorphisms between subsets in pA of local transversals to projected phase curves and open subsets of the attractor $A(P)$ of the return map P.

In Chapter 9 we investigate angles along projections of phase curves of slowly oscillating solutions. As long as a slowly oscillating solution x has zeros, the projected phase curve $t \mapsto px_t$ winds around the origin. However, the angle of the vector px_t (with respect to 0 and a reference line in L) does not necessarily increase monotonically in the course of this. We refine homotopy techniques from [20] and obtain lower estimates of the angle in terms of an *increasing* function.

These a–priori estimates are essential for our proof of the Poincaré–Bendixson results in Chapter 10.

The final Chapter 11 begins with the construction of a periodic orbit in A whose projection passes through a point in pA with maximal norm. The exterior of the projected periodic orbit and pA are disjoint. It remains to exclude a hole of pA in the interior. This is done using the connectedness of the attractors A and $A(P)$.

All results and proofs are formulated for equation (1.1) with $\mu > 0$. The versions for the case $\mu = 0$ are simpler and can easily be extracted.

A question left open is whether the attractor A is C^1–smooth. In [19, 20] we obtained smooth submanifolds whose closures coincide with A in special cases (compare e.g. [21]).

It may be asked what happens if the monotonicity assumption on f is dropped. In [10] we show that there exist smooth, bounded functions f which are not monotone, but satisfy $\xi f(\xi) < 0$ for all $\xi \neq 0$, and define an attractor $A \subset \overline{S}$ for eq. (1.1) which contains a normally hyperbolic invariant set on which the dynamics are chaotic.

In the sequel, a few general results are used without comment. As references we recommend [5, 2, 19] for facts about differential delay equations, [3] for calculus in Banach spaces and winding numbers, remarks in [1] for angles along curves in the plane, and [6] for attractors and related objects.

CHAPTER 2

Notation, Preliminaries

$\mathbb{N}_0$ denotes the set of nonnegative integers. $\mathbb{R}^+$ stands for the interval $[0, \infty)$. Let E be a real or complex Banach space. E^* denotes the Banach space of continuous linear functionals on E. A curve in E is a continuous map $c : J \to E$, defined on an interval $J \subset \mathbb{R}$. The image $|c| = c(J)$ is called the trace of c. A curve is said to be (piecewise) smooth if it is (piecewise) of class C^1. Let a subset $M \subset E$ be given. The closure of M is denoted by $\overline{M}$, its boundary by ∂M. The tangent set $T_x M$ of M at a point x of M is the set of all vectors

$$v = Dc(0)1$$

where $c : (-1, 1) \to E$ is a differentiable curve with $c(0) = x$ and $|c| \subset M$.

For a piecewise smooth curve $c : [a, b] \to \mathbb{C}$ with $0 \notin |c|$ and for $\omega_0 \in \mathbb{R}$ with $c(a) = |c(a)|e^{i\omega_0}$ we define the angle along c by

$$\omega(t) = \Re \frac{1}{i} \int_{c|[a,t]} \frac{1}{z} dz + \omega_0, \quad a \leq t \leq b.$$

The function ω is continuous.

LEMMA 2.1. $c(t) = |c(t)|e^{i\omega(t)}$ for $a \leq t \leq b$.

PROOF. We only consider the case that c is smooth. Using a local inverse of the polar coordinate map

$$(r, \gamma) \mapsto r(\cos \gamma, \sin \gamma)$$

we obtain $\epsilon > 0$ and a C^1–map

$$\beta : [a, a + \epsilon) \ni t \mapsto (\Re c(t), \Im c(t)) \mapsto (|c(t)|, \beta(t)) \mapsto \beta(t) \in \mathbb{R};$$

for $a \leq t < a + \epsilon$,

$$c(t) = |c(t)|e^{i\beta(t)}, \quad \text{and} \quad \beta(a) = \omega_0.$$

It follows that for such t

$$
\begin{aligned}
\omega(t) &= \Re \frac{1}{i} \int_{c|[a,t]} \frac{1}{z} dz + \omega_0 \\
&= \Re \frac{1}{i} \int_a^t \frac{c'(s)}{c(s)} ds + \omega_0 \\
&= \Re \frac{1}{i} \int_a^t \frac{1}{c(s)} \frac{d|c|}{ds}(s) ds + \int_a^t \beta'(s) ds + \beta(a) \\
&= \beta(t)
\end{aligned}
$$

Hence

$$
c(t) = |c(t)| e^{i\omega(t)} \quad \text{for} \quad a \leq t < a + \epsilon.
$$

Suppose

$$
c(s) \neq |c(s)| e^{i\omega(s)} \quad \text{for some} \quad s \in [a + \epsilon, b].
$$

Then there exists $t_0 \in [a + \epsilon, b]$ such that on $[a, t_0]$,

$$
c(t) = |c(t)| e^{i\omega(t)},
$$

and for a sequence $t_n \searrow t_0$,

$$
c(t_n) \neq |c(t_n)| e^{i\omega(t_n)}.
$$

The argument from above, now applied to $c|[t_0, b]$ and $\omega(t_0)$ instead of c and ω_0, yields $\eta > 0$ such that for $t_0 \leq t < t_0 + \eta$,

$$
\begin{aligned}
\Re \frac{1}{i} \int_{c|[t_0,t]} \frac{1}{z} dz + \omega(t_0) &= \\
\cdots + \Re \frac{1}{i} \int_{c|[a,t_0]} \frac{1}{z} dz + \omega_0 &= \\
&= \omega(t)
\end{aligned}
$$

satisfies

$$
c(t) = |c(t)| e^{i\omega(t)},
$$

a contradiction. □

COROLLARY 2.1. *Suppose $|c| \subset V$ for a convex cone $V \subset \mathbb{C} \setminus \{0\}$. Then $|\omega(b) - \omega(b)| < \pi$.*

PROOF. Otherwise there would exist $t \in (a, b]$ such that $|\omega(t) - \omega(a)| = \pi$; and the straight line segment from $c(a)$ to $c(t)$ would contain 0 which contradicts the assumptions about V. □

COROLLARY 2.2. *In case*

$$
c(t) = \frac{1}{b - a}((t - a)c(b) + (b - t)c(a)) \neq 0 \quad \text{for} \quad a \leq t \leq b,
$$

$$
|\omega(b) - \omega(a)| < \pi.
$$

Recall that for a closed piecewise smooth curve c and for a point $y \in \mathbb{C} \setminus |c|$ the winding number is defined by

$$\text{wind}(y, c) = \frac{\omega(b) - \omega(a)}{2\pi}$$

where ω is an angle along $c - y$. We shall make use of the following facts:

$$\text{wind}(y, c) \in \mathbb{Z};$$

if c is a simple closed piecewise smooth curve then

$$\text{wind}(y, c) \in \{-1, 0, 1\}.$$

For such c the interior and the exterior regions are defined by

$$\text{int}(c) = \{y \in \mathbb{C} \setminus |c| : \text{wind}(y, c) \neq 0\},$$

$$\text{ext}(c) = \{y \in \mathbb{C} \setminus |c| : \text{wind}(y, c) = 0\}.$$

Both sets are open and connected. $\text{int}(c)$ is bounded, $\text{ext}(c)$ is unbounded, and the boundaries coincide with the trace $|c|$. The latter implies that in case c is smooth and $c'(t) \notin \mathbb{R}d$ for some $t \in [a, b]$ and $d \neq 0$, we have either

$$c(t) + (-r, 0)d \subset \text{int}(c) \quad \text{and} \quad c(t) + (0, r)d \subset \text{ext}(c) \quad \text{for some} \quad r > 0,$$

or

$$c(t) + (-r, 0)d \subset \text{ext}(c) \quad \text{and} \quad c(t) + (0, r)d \subset \text{int}(c) \quad \text{for some} \quad r > 0.$$

The winding number is a homotopy invariant: If $h : [0, 1] \times [a, b] \to \mathbb{C}$ is continuous, if

$$y \in \mathbb{C} \setminus h([0, 1] \times [a, b]), \quad h(\cdot, a) = h(\cdot, b),$$

and if both $h(0, \cdot)$ and $h(1, \cdot)$ are piecewise smooth, then

$$\text{wind}(y, h(0, \cdot)) = \text{wind}(y, h(1, \cdot)).$$

(It is not necessary that the closed curves $h(t, \cdot), 0 < t < 1$, are piecewise smooth.)

If Y is any 2–dimensional $\mathbb{R}$–vectorspace and if I is an isomorphism onto the $\mathbb{R}$–vectorspace $\mathbb{C}$, then we define angles along a piecewise smooth curve c in Y by the angles along $I \circ c$. Analogously, we define for simple closed piecewise smooth curves c in Y and points $y \in Y \setminus |c|$

$$\text{wind}(y, c) = \text{wind}(I(y), I \circ c), \quad \text{int}(c) = \text{int}(I \circ c), \quad \text{ext}(c) = \text{ext}(I \circ c).$$

Let a function $g : \mathbb{R}^2 \to \mathbb{R}$ be given. A solution of the differential delay equation

(2.1)
$$x'(t) = g(x(t), x(t - 1))$$

is either a differentiable function $x : \mathbb{R} \to \mathbb{R}$ so that (2.1) is satisfied for all real t, or a continuous function $x : [t_0 - 1, \infty) \to \mathbb{R}$, $t_0 \in \mathbb{R}$, which is differentiable on (t_0, ∞) and satisfies (2.1) for all $t > t_0$.

Analogously one defines complex-valued solutions in case g is linear, and solutions of nonautonomous equations

$$x'(t) = g(t, x(t-1))$$

for functions $g : \mathbb{R}^2 \to \mathbb{R}$ or $g : [t_0, \infty) \times \mathbb{R} \to \mathbb{R}, t_0 \in \mathbb{R}$.

The space C, and the complex vectorspace C' of continuous functions ϕ : $[-1, 0] \to \mathbb{C}$, are equipped with the maximum–norm:

$$\|\phi\| = \max_{t \in [-1,0]} |\phi(t)|$$

Solutions x define phase curves $X : t \mapsto x_t$ with values in C or C' by

$$X(t)(s) = x_t(s) = x(t+s) \quad \text{for all} \quad s \in [-1, 0],$$

provided the interval $[t-1, t]$ belongs to the domain of x.

The term trajectory is used in connection with maps: If P is a mapping, then sequences (ϕ_n) with $\phi_{n+1} = P(\phi_n)$ are called trajectories.

PROPOSITION 2.1. *Let* v, s, w, u *in* $\mathbb{C}$ *and* $\epsilon > 0$ *be given. If*

(2.2)
$$|v - s| = \epsilon,$$

(2.3)
$$|w - u| = \epsilon,$$

(2.4)
$$|v - u| \geq \epsilon,$$

(2.5)
$$|w - s| \geq \epsilon,$$

$$u \neq s,$$

$$\emptyset \neq (v + [0, 1](s - v)) \cap (w + [0, 1](u - w)),$$

then

$$v = w.$$

PROOF. 1. Choose $z \in (v + [0, 1](s - v)) \cap (w + [0, 1](u - w))$. Claim: $z \neq s$. Proof. Suppose $z = s$. Then $s \in w + [0, 1](u - w)$. (2.3) and (2.5) imply $s = u$, a contradiction.

2. It follows that $z = v + t(s - v)$ with $0 \leq t < 1$;

$$v = s + \frac{1}{1-t}(z - s).$$

Set $r = \frac{1}{1-t} \geq 1$.

3. Claim: $|w - s| = \epsilon$ and $|w - s| = |w - z| + |z - s|$. Proof. We have

$$
\begin{aligned}
2\epsilon \;&\leq\; |w - s| + |v - u| \\
&\leq\; |w - z| + |z - s| + |v - z| + |z - u| \\
&=\; (|w - z| + |z - u|) + (|v - z| + |z - s|) \\
&=\; |w - u| + |v - s| \quad \text{(see 1.)} \\
&=\; 2\epsilon.
\end{aligned}
$$

Using (2.4) and (2.5) we infer

$$
|w - s| = \epsilon.
$$

Also, the assumption $|w - z| + |z - s| > \epsilon$ would imply

$$
2\epsilon < |w - z| + |z - s| + |v - u| \leq \cdots = 2\epsilon,
$$

a contradiction.

4. Claim: For some $\rho > 0, w - s = \rho(z - s)$. Proof. In case $w \neq z$ we infer from 3. that the nonzero vectors $w - z, z - s$ are linearly dependent elements of the $\mathbb{R}$–vectorspace $\mathbb{C}$, and that they are positive multiples of each other. Use $w - s = (w - z) + (z - s)$.

5. From 2., (2.2), 3., and 4., we get $\rho = r$, and $v = w$. $\square$

We shall make use of several results obtained in [20]. So it might be convenient here to correct errors in [20]:

Page 181, line 27: Read "is injective". Delete "$D_2 F$ is continuous on $\mathbb{R} \times C$."

Page 183, lines 23 and 24: Read

$$
\begin{aligned}
g(t,\xi) \;&=\; e^{\mu t}[f(e^{-\mu(t-1)}\xi + x^\psi(t - 1)) - f(x^\psi(t - 1))] \\
&=\; e^{\mu t} \int_{x^\psi(t-1)}^{e^{-\mu(t-1)}\xi + x^\psi(t-1)} f'.
\end{aligned}
$$

Page 204, line 6: Replace "$\phi \in N_0$" by "$\|\phi\| \leq b_f$".

Page 204, lines 7 and 8: Replace these lines by the following: "(In order to derive the last estimate, show that the set

$$
\{x^\phi(s) : \|\phi\| \leq b_f, -1 \leq s \leq \tau\}
$$

is bounded, and work with equation (3.1).)"

Page 204, line 21: Read "ϕ_0" instead of "ϕ_j".

Page 237, line 11: Delete "newline".

CHAPTER 3

Basic Properties of Solutions

Let a continuously differentiable function $f : \mathbb{R} \to \mathbb{R}$ be given with the properties

$$f(0) = 0, \quad f'(\xi) < 0 \quad \text{for all} \quad \xi, \quad \sup f < \infty.$$

Let a constant $\mu > 0$ be given.

Every $\phi \in C$ determines a solution $x^\phi : [-1, \infty) \to \mathbb{R}$ of equation (1.1) with $x_0^\phi = \phi$. This is most easily seen from repeated applications of the variation–of–constants formula

$$x(t) = e^{-\mu t} x(n-1) + \int_{n-1}^t e^{-\mu(t-s)} f(x(s-1)) ds$$

to the intervals $[n-1, n], n \in \mathbb{N}$. Another useful observation is that for every solution x of equation (1.1) the function $\hat{x} : t \mapsto e^{\mu t} x(t)$ is a solution of the equation

(3.1) $$\hat{x}'(t) = g(t, \hat{x}(t-1))$$

with $g(t, \xi) = e^{\mu t} f(e^{-\mu(t-1)} \xi)$. Note that

(3.2) $$g(t, 0) = 0 \quad \text{and} \quad g(t, \xi)\xi < 0 \quad \text{for all} \quad t \quad \text{and} \quad \xi \neq 0.$$

We collect a series of basic facts about solutions of equation (1.1). For proofs, see e.g. [19].

The solutions x^ϕ define a continuous semiflow

$$F : \mathbb{R}^+ \times C \ni (t, \phi) \to x_t^\phi \in C.$$

We have continuous dependence on initial data also in the following sense:

For $\phi \in C, t \geq 0$, and $\epsilon > 0$ given, there exists $\delta > 0$ such that for all $\psi \in C$ with $|\psi - \phi| \leq \delta$ and for all $s \in [0, t]$,

$$|x^\psi(s) - x^\phi(s)| < \epsilon.$$

The Arzela–Ascoli theorem, applied to the variation–of–constants formula for $x^\phi(t), 0 \leq t \leq 1$, yields that the map $F(1, \cdot)$ is compact in the sense that it

11

maps bounded sets into sets with compact closure. It follows that all maps $F(t, \cdot), t \geq 1$, are compact.

F is of class C^1 on $(1, \infty) \times C$; each $F(t, \cdot), t \geq 0$, is of class C^1. For $t > 1, \phi \in C$, and $x = x^\phi$, we have

$$D_1 F(t, \phi)1 = (x')_t.$$

In the sequel we write x'_t instead of $(x')_t$. For $t \geq 0, \phi \in C, \psi \in C$,

$$D_2 F(t, \phi)\psi = v_t$$

where $v : [-1, \infty) \to \mathbb{R}$ is the solution of the *linear variational equation along* $x = x^\phi$,

$$v'(t) = -\mu v(t) + f'(x(t-1))v(t-1),$$

which satisfies $v_0 = \psi$. If $x : [-2, \infty) \to \mathbb{R}$ is a solution of equation (1.1) then

$$D_2 F(l, x_0)x'_0 = x'_t \quad \text{for all} \quad t \geq 0.$$

PROPOSITION 3.1.   (i) *If* $x : [t_0 - 1, \infty) \to \mathbb{R}$ *is a solution of equation* (1.1) *such that* x *has no zero on* $[t_0, \infty)$ *and for* $t < t_0$, *either* $x(t) = 0$ *or* $\mathrm{sign}(x(t)) = \mathrm{sign}(x(t_0))$, *then* $\mathrm{sign}(x'(t)) = -\mathrm{sign}(x(t))$ *for* $t > t_0$ *and* $x(t) \to 0$ *as* $t \to \infty$.

(ii) *If* $x : \mathbb{R} \to \mathbb{R}$ *is a solution which is bounded on* $(-\infty, 0]$ *then*

$$\inf x^{-1}(0) = -\infty.$$

(iii) *There exists a constant* $b(\mu, f) > 0$ *such that for all* $\phi \in C$,

$$\limsup_{t \to \infty} \|F(t, \phi)\| \leq b(\mu, f).$$

(iv) *There exists a compact subset* $C_{\mu, f} \subset C$ *such that for every* $\phi \in C$ *there is* $t \geq 0$ *with* $F(s, \phi) \in C_{\mu, f}$ *for all* $s \geq t$, *and for every bounded solution* $x : \mathbb{R} \to \mathbb{R}$ *of equation* (1.1), $x_t \in C_{\mu, f}$ *for all* $t \in \mathbb{R}$.

PROOF. 1. Proof of (i). Suppose $x(t) > 0$ for $t \geq t_0$, and $x(t) \geq 0$ for $t_0 - 1 \leq t < t_0$. Equation (1.1) implies that for $t > t_0, x'(t) < 0$. It follows that $x(t)$ converges to some $c \geq 0$ as $t \to \infty$. Suppose $c > 0$. By equation (1.1), $x'(t) \to -\mu c + f(c) < 0$ as $t \to \infty$. This implies a contradiction. The other case is analogous.

2. Proof of (ii). Suppose $x(t) \neq 0$ for $t \leq t_0$. The solution $\hat{x} : t \mapsto e^{\mu t}x(t)$ of equation (3.1) satisfies $\mathrm{sign}(\hat{x}(t)) = -\mathrm{sign}(\hat{x}'(t)) \neq 0$ for $t \leq t_0$ and $\hat{x}(t) \to 0$ as $t \to -\infty$. This implies a contradiction.

3. Proof of (iii). Assertion 1 implies that solutions x^ϕ with a bounded zeroset decay to 0 as $t \to \infty$; for such $\phi, \lim_{t \to \infty} F(t, \phi) = 0$. Suppose the zeroset of $x = x^\phi$ is unbounded. For every local maximum $m > 1$, we have $0 = x'(m) = -\mu x(m) + f(x(m-1))$ and $x(m) > 0$. Therefore $x(m-1) < 0$, and there exists a largest zero $z \in (m-1, m)$ of x;

$$0 < x(m) = \int_z^m x'(t)dt \leq \int_z^m f(x(t-1))dt \leq \sup f < \infty.$$

Fix a zero $z > 1$ of x. We saw that $x(t) \leq \sup f$ for $t \geq z$. Analogous arguments imply that for every local minimum $m > z + 1$,

$$0 > x(m) \geq \min\{f(\xi) : 0 \leq \xi \leq \sup f\}.$$

Set $b(\mu, f) = \sup f + \min\{f(\xi) : 0 \leq \xi \leq \sup f\}$. Choose a zero $\bar{z} > z + 1$. Then $|x(t)| \leq b(\mu, f)$ for $t \geq \bar{z}$. If $x : \mathbb{R} \to \mathbb{R}$ is a bounded solution of equation (1.1), then $x^{-1}(0)$ has no lower bound, and arguments as above show that

$$|x(t)| \leq b(\mu, f) \quad \text{for all} \quad t \in \mathbb{R}.$$

4. In order to prove (iv) set $C_{\mu, f} = \overline{\{F(1, \phi) : \|\phi\| \leq b(\mu, f)\}}$. $\square$

The fact that a compact subset attracts each point $\phi \in C$ implies that F is point dissipative in the sense of [6, p. 38].

Furthermore, every solution $x^{\phi} : [-1, \infty) \to \mathbb{R}$ has a nonempty ω–limit set

$$\omega(\phi) = \{\psi \in C : \text{There is a sequence } (t_n)_0^{\infty} \text{ such that }$$
$$t_n \to \infty, F(t_n, \phi) \to \psi\},$$

which is compact and connected. For each $\psi \in \omega(\phi)$ there is a solution $\bar{x} : \mathbb{R} \to \mathbb{R}$ of equation (1.1) so that $\bar{x}_0 = \psi$ and $\bar{x}_t \in \omega(\phi)$ for all $t \in \mathbb{R}$. Likewise, every bounded solution $x : \mathbb{R} \to \mathbb{R}$ of equation (1.1) has a nonempty α–limit set

$$\alpha(x) = \{\psi \in C : \text{There is a sequence } (t_n)_0^{\infty} \text{ such that }$$
$$t_n \to -\infty, X(t_n) \to \psi\},$$

which is compact and connected. For each $\psi \in \alpha(x)$ there exists a solution $\bar{x} : \mathbb{R} \to \mathbb{R}$ of equation (1.1) with $\bar{x}_0 = \psi$ and $\bar{x}_t \in \alpha(x)$ for all $t \in \mathbb{R}$.

An important consequence of the strict monotonicity of f is that all maps $F(t, \cdot), t \geq 0$, are injective. (This follows easily by means of the variation–of–constants formula for solutions.) In particular, any two solutions $x : \mathbb{R} \to \mathbb{R}, y : \mathbb{R} \to \mathbb{R}$ with $x_t = y_t$ for some $t \in \mathbb{R}$ coincide. Therefore we may sharpen the statements about phase curves through points in ω– and α–limit sets: For $\phi \in C$ given, each $\psi \in \omega(\phi)$ determines a unique solution $x^*(\psi) : \mathbb{R} \to \mathbb{R}$ of equation (1.1) such that $x^*(\psi)_0 = \psi$ and $x^*(\psi)_t \in \omega(\phi)$ for all $t \in \mathbb{R}$. If $x : \mathbb{R} \to \mathbb{R}$ is a bounded solution of equation (1.1) then each $\psi \in \alpha(x)$ determines a unique solution $x^*(\psi) : \mathbb{R} \to \mathbb{R}$ of equation (1.1) such that $x^*(\psi)_0 = \psi$ and $x^*(\psi)_t \in \alpha(x)$ for all $t \in \mathbb{R}$.

A solution is called *slowly oscillating* (with respect to the delay 1) if for every pair of zeros $z' > z$,

$$z' - z > 1.$$

It is clear that segments x_t of slowly oscillating solutions belong to the set S of nonzero functions with at most one change of sign; i.e.

$$\phi \in S \quad \text{if and only if} \quad \phi \neq 0, \quad \text{and there exists}$$

$$z \in [-1, 0] \quad \text{such that} \quad \phi \leq 0 \quad \text{in} \quad [-1, z], \quad 0 \leq \phi \quad \text{in} \quad [z, 0],$$

or there is $z \in [-1, 0]$ with $0 \leq \phi$ in $[-1, z]$, $\phi \leq 0$ in $[z, 0]$.

S is a wedge (if $t > 0$ and $\phi \in S$, then $t\phi \in S$), and not convex. One can show that S is homotopy equivalent to a circle [4]. Elementary considerations yield

$$\overline{S} = S \cup \{0\}.$$

The next results show in particular that slowly oscillating solutions are abundant. Set

$$K = \{\phi \in C : \phi(-1) = 0, \quad t \mapsto e^{\mu t}\phi(t) \text{ is increasing}, \quad 0 < \phi(0)\}.$$

The set K is a convex cone contained in S.

PROPOSITION 3.2.   (i) *Let $\phi \in C$ with $\phi \geq 0, \phi \neq 0$ be given. Consider $x = x^\phi$. Either $x(t) > 0$ for all $t \geq 0$, and $x(t) \searrow 0$ as $0 \leq t \to \infty$, or there is a smallest zero z of x in $\mathbb{R}^+$, and $x_{z+1} \in -K$.*

(ii) *Let $\phi \in K$. Then the restriction of $x = x^\phi$ to $\mathbb{R}^+$ is slowly oscillating. If $z > 0$ is a zero of x then $|x|$ and $|x'|$ are bounded on the interval $[z, z+1]$ by*

$$(1 + \mu) \cdot \max\{|f(\xi)| : \xi \in x([z - 1, z])\},$$

and the function $t \mapsto e^{\mu t}|x_{z+1}(t)|$ is increasing.
If the zeroset of $x|\mathbb{R}^+$ is unbounded then it is given by a sequence of points $z_j = z_j(\phi), j \in \mathbb{N}$, with

(so) $z_j + 1 < z_{j+1}$ and $x'(z_j) \neq 0$ for all j;

and x is monotone on $[0, z_1]$ and on each interval $[z_j + 1, z_{j+1}]$.
If the zeroset of $x|\mathbb{R}^+$ is bounded then it is given by a finite sequence of points $z_j = z_j(\phi), 1 \leq j \leq J = J(\phi)$, with property (so), and $x(t)$ decreases monotonically to 0 on the interval $[z_J + 1, \infty)$ as $t \to \infty$.

PROOF. 1. Proof of (i). In case $x(t) > 0$ for $t > 0$, apply Proposition 3.1. If x has a zero on $\mathbb{R}^+$, set $z = \min x^{-1}(0) \cap \mathbb{R}^+$. The solution $\hat{x} : t \mapsto e^{\mu t}x(t)$ of equation (3.1) satisfies $\hat{x}(t) \geq 0$ for $t \leq z$, and $x(s) > 0$ for some $s \in [z - 1, z)$. Therefore, property (3.2) implies that $\hat{x}' \leq 0$ on $[z, z + 1)$, and $\hat{x}'(s) < 0$; it follows that $x_{z+1} \in -K$.

2. Proof of (ii). Consider $\hat{x}$ as before. There exists $z \in [-1, 0]$ with $\hat{x}(t) = 0$ for $t \leq z$ and $0 < \hat{x}(t)$ for $z < t \leq 0$. Equation (3.1) and property (3.2) imply $\hat{x}(t) = \hat{x}(0) > 0$ for $0 \leq t \leq z + 1$ and $\hat{x}'(t) < 0$ for $z + 1 < t \leq 1$. In case $\hat{x}(t) > 0$ for all $t > 0$, x is slowly oscillating. Otherwise there exists a smallest zero z_1 of $\hat{x}$ (and of x) in $(z + 1, \infty)$. Equation (3.1) and property (3.2) give $\hat{x}'(t) < 0$ for $z \leq t < z + 1$. In particular, $X(z_1 + 1) \in -K$. Iterate the previous argument. This yields the assertions about the zeroset of x. Proposition 3.1 and the argument in part 1 of its proof imply the statements about monotonicity.

Let a zero $z > 0$ of x be given. Suppose $x'(z) < 0$. Then $x(s) < 0$ for $z < s \leq z + 1$. Let $z \leq t \leq z + 1$. In case $0 > x'(t)$ we get $x'(t) \geq f(x(t-1))$. This implies

$$0 \geq x(s) = \int_z^s x'(t)dt \geq \max\{|f(\xi)| : \xi \in x([z-1,z])\}$$

for $z \leq s \leq z + 1$. In case $0 < x'(t)$, we conclude from the latter (since $f(x(t-1)) \leq 0$) that

$$x'(t) \leq -\mu x(t) \leq \mu \max\{|f(\xi)| : \xi \in x([z-1,z])\}.$$

The proof for $x'(z) > 0$ is analogous. $\square$

The set S is positively invariant:

$$F(\mathbb{R}^+ \times S) \subset S,$$

and, more generally than above, every $\phi \in S$ defines a solution $x = x^\phi$ such that for t_0 sufficiently large, $x|[t_0, \infty)$ is slowly oscillating.

The monotonicity of f implies an even stronger statement.

PROPOSITION 3.3. *For initial data ϕ, ψ with $\phi - \psi \in S$, there exists $s \in [0,4]$ such that $F(s, \phi) - F(s, \psi)$ has no zero;*

$$F(t, \phi) - F(t, \psi) \in S \quad \text{for all} \quad t \geq 0.$$

PROOF. Apply [**19**, Remark 6.1] and [**19**, Proposition 6.1] to the solution

$$t \mapsto e^{\mu t}(x^\phi(t) - x^\psi(t))$$

of the equation

$$x'(t) = g(t, x(t-1))$$

where

$$g(t, \xi) = e^{\mu t}[f(e^{-\mu(t-1)}\xi + x^\psi(t-1)) - f(x^\psi(t-1))]$$

$$= e^{\mu t} \int_{x^\psi(t-1)}^{e^{-\mu(t-1)}\xi + x^\psi(t-1)} f'.$$

$\square$

It follows that every bounded solution $x : \mathbb{R} \to \mathbb{R}$ of equation (1.1) with $x_t \in S$ for all t is slowly oscillating, and that the zeroset of x is given by a sequence of points $z_j = z_j(x), j \in \mathbb{Z}$, or $j \in \mathbb{Z}$ and $j \leq J = J(x)$ for some $J \in \mathbb{Z}$, so that property (so) holds. In case the zeroset of such a solution is not bounded from above, set $J(x) = \infty$.

We shall make use of a return map on the closed convex cone

$$\overline{K} = K \cup \{0\} \subset \overline{S}.$$

If $\phi \in K$ and if the zeroset of x^ϕ is unbounded, set $J(\phi) = \infty$. Proposition 3.2 permits to define a map $P : \overline{K} \to \overline{K}$ by

$$P(\phi) \;=\; F(z_2(\phi)+1,\phi) \quad \text{if} \quad \phi \in K, \quad \text{if} \quad x^\phi \quad \text{has zeros in}$$
$$\mathbb{R}^+, \quad \text{and if} \quad J(\phi) \geq 2,$$
$$P(\phi) \;=\; 0 \quad \text{otherwise.}$$

PROPOSITION 3.4. *P is continuous, and $P(\overline{K})$ has compact closure.*

PROOF. 1. Set $x = x^\phi$, for $\phi \in \overline{K}$. If $\phi \neq 0$ and if x has zeros on $\mathbb{R}^+$, set $J = J(\phi)$ and $z_j = z_j(\phi)$.

2. Continuity at $\phi = 0$ and at $\phi \in K$ with $J \geq 2$ follows as in the proof of [**19**, Proposition 9.2].

3. The case $\phi \in K$ and $J = 1$. Proposition 3.2(ii) gives $x(t) \to 0$ as $t \to \infty$. Let $\epsilon > 0$. There exists $t > z_1 + 3$ with $\|x_t\| < \epsilon$. By continuous dependence on initial data there is a neighbourhood N of ϕ in $\overline{K}$ such that for every solution $y : [-1,\infty) \to \mathbb{R}$ with $y_0 = \psi \in N$,

$$y(s) > 0 \quad \text{for} \quad 0 \leq s \leq z_1 - \frac{1}{2}, \quad y(s) < 0 \quad \text{for} \quad z_1 + \frac{1}{2} \leq s, \quad \|y_t\| < 2\epsilon.$$

Hence $z_1(y) < z_1 + \frac{1}{2}$. In case $J(\psi) = 1$, we have $P(\psi) = 0 = P(\phi)$. In case $J(\psi) \geq 2$, we infer $z_2(\psi) > t$. The assertion in Proposition 3.2(ii) on monotonicity yields $|y(s)| < 2\epsilon$ for $z_2(\psi) - 1 \leq s \leq z_2(\psi)$ since $z_1(y) + 1 < t - 1 < t < z_2(y)$. Furthermore,

$$\|P(\psi)\| = \|F(z_2(\psi)+1,\psi)\| \leq (\mu+1)\max_{[-2\epsilon,2\epsilon]} |f|.$$

4. In case x has no zeros on $\mathbb{R}^+$, consider, for $\epsilon > 0$ given, $t > 1$ and a neighbourhood N of ϕ in $\overline{K}$ such that for every solution $y : [-1,\infty) \to \mathbb{R}$ with $y_0 = \psi \in N$,

$$y(s) > 0 \quad \text{for} \quad 0 \leq s \leq t \quad \text{and} \quad \|y_t\| < \epsilon.$$

In case y has at most one zero in $\mathbb{R}^+$, $P(\psi) = 0 = P(\phi)$. In case $J(\psi) \geq 2$, set $\eta = (\mu+1)\max_{[-2\epsilon,2\epsilon]} |f|$ and show successively that

$$|y(s)| \leq \eta \quad \text{for} \quad z_1(\psi) \leq s \leq z_1(\psi)+1,$$

$$|y(s)| \leq \eta \quad \text{for} \quad z_2(\psi) - 1 \leq s \leq z_2(\psi),$$

and finally,

$$|y(s)| \leq (\mu+1)\max_{[-\eta,\eta]} |f| \quad \text{for} \quad z_2(\psi) \leq s \leq z_2(\psi)+1.$$

5. Compactness. In case $J \geq 2$, we have $x(t) < 0$ for $z_2 - 1 \leq t < z_2$. Proposition 3.2(ii) now implies that both $|x|$ and $|x'|$ are bounded on $[z_2, z_2+1]$ by $(\mu+1)\sup f$. Apply the theorem of Arzela–Ascoli. $\square$

Observe that

$$x_{z+1} \in K$$

whenever z is a zero of a bounded solution $x : \mathbb{R} \to \mathbb{R}$ with phase curve in S and $x'(z) > 0$.

The next result will imply that close to points $\phi \in K$ such that x^ϕ has zeros on $\mathbb{R}^+$ and $J(\phi) \geq 2$, the return map P is given by a continuously differentiable intersection map. Consider the closed hyperplane

$$H = \{\phi \in C : \phi(-1) = 0\} = \ker(ev)$$

where ev denotes the evaluation functional $\phi \mapsto \phi(-1)$. Note that $\overline{K} \subset H$. Each phase curve $t \mapsto x_t$ of a solution x which starts at some $x_0 = \phi \in K$ and has at least 2 zeros on $\mathbb{R}^+$ intersects transversally with H at $t = z_2 + 1, z_2 = z_2(\phi)$, since

$$0 \neq x'(z_2 + 1) = x_t'(-1) = [D_1(s \mapsto F(s, \phi))(t)](-1) = ev([\ldots]).$$

PROPOSITION 3.5. *Let $\phi \in C$ and $t > 1$ be given such that $x = x^\phi$ satisfies $x(t-1) = 0$ and $x'(t-1) \neq 0$. Then there exist an open neighbourhood U of ϕ in $C, \epsilon > 0$, and a continuously differentiable map $\sigma : U \to (t - \epsilon, t + \epsilon)$ such that $\sigma(\phi) = t$, and for $(s, \psi) \in (t - \epsilon, t + \epsilon) \times U$,*

$$F(s, \psi) \in H \quad \text{if and only if} \quad s = \sigma(\psi).$$

PROOF. Solve the equation $ev \circ F(s, \psi) = 0$ in a neighbourhood of the point (t, ϕ) by means of the Implicit Function Theorem. $\square$

The map $U \ni \psi \mapsto F(\sigma(\psi), \psi) \in C$ is continuously differentiable.

COROLLARY 3.1. *For every $\phi \in K$ such that $x = x^\phi$ has zeros on $\mathbb{R}^+$ and $J(\phi) \geq 2$ there exist a neighbourhood V of ϕ in $C, \delta > 0$, and a continuously differentiable map $\sigma : V \to (z_2(\phi) + 1 - \delta, z_2(\phi) + 1 + \delta)$ such that for every $\psi \in V \cap K, x^\psi$ has zeros on $\mathbb{R}^+, J(\psi) \geq 2, \sigma(\psi) = z_2(\psi) + 1$, and $P(\psi) = F(\sigma(\psi), \psi)$.*

PROOF. Continuous dependence on initial data permits to find a neighbourhood N of ϕ in K such that for all $\psi \in N$ and $y = x^\psi$, we have $y(s) > 0$ for $0 \leq s \leq \max\{0, z_1(\phi) - \frac{1}{2}\}, y(s) < 0$ for $z_1(\phi) + \frac{1}{2} \leq s \leq z_2(\phi) - \frac{1}{2}$, and $y(z_2(\phi) + \frac{1}{2}) > 0$. Therefore y has zeros on $\mathbb{R}^+, J(\psi) \geq 2$, and

$$z_1(\phi) - \frac{1}{2} < z_1(\psi) < z_1(\phi) + \frac{1}{2},$$
$$z_2(\phi) - \frac{1}{2} < z_2(\psi) < z_2(\phi) + \frac{1}{2}.$$

Note that y has exactly one zero in the interval $(z_1(\phi) - \frac{1}{2}, z_1(\phi) + \frac{1}{2})$, and exactly one zero in the interval $(z_2(\phi) - \frac{1}{2}, z_2(\phi) + \frac{1}{2})$. Now apply Proposition 3.5 to ϕ, with $t = z_2(\phi) + 1$. Set $\delta = \min\{\epsilon, \frac{1}{2}\}$, and choose a neighbourhood $V \subset U$ of ϕ in C so small that $V \cap K \subset N$ and $\sigma(V) \subset (z_2(\phi) + 1 - \delta, z_2(\phi) + 1 + \delta)$. Conclude that $\sigma(\psi) = z_2(\psi) + 1$. $\square$

CHAPTER 4

Attractors

Consider a complete metric space M, a semiflow $G : \mathbb{R}^+ \times M \to M$, and a subset $N \subset M$. The set N is called invariant if for each $u_0 \in N$ there exists a phase curve $u : \mathbb{R} \to M$ (i.e., $u(t) = G(t - s, u(s))$ for $t \geq s$) with $u_0 = u(0)$ and $u(\mathbb{R}) \subset N$, and if every phase curve $u : \mathbb{R} \to M$ with $u(0) = u_0$ satisfies $u(\mathbb{R}) \subset N$. The set N is called maximal compact invariant if it is compact and invariant and if every compact invariant set N' is contained in N. The set is said to attract a set $B \subset M$ if for every open set $U \subset M$ with $N \subset U$ there exists $t \geq 0$ such that $\{G(s, b) : b \in B\} \subset U$ for all $s \geq t$. A global attractor is a maximal compact invariant set which attracts every bounded set.

[6, Theorem 3.4.8] implies that the semiflow F has a global attractor $A(F)$. The restriction of F to $\mathbb{R}^+ \times \overline{S}$ defines a semiflow F_S on the complete metric space $\overline{S}$ which has a global attractor

$$A = A(F_S) \subset \overline{S}.$$

PROPOSITION 4.1. *Let $\phi \in C$ be given. Then $\phi \in A$ if and only if there exist $t \in \mathbb{R}$ and a bounded solution $x : \mathbb{R} \to \mathbb{R}$ of equation (1.1) with $x_t = \phi$ and $x_s \in \overline{S}$ for all $s \in \mathbb{R}$.*

PROOF. For every bounded solution $x : \mathbb{R} \to \mathbb{R}$ with phase curve in $\overline{S}$ the bounded set $B = \{x_t : t \in \mathbb{R}\} \subset \overline{S}$ satisfies $\{F(t, \phi) : \phi \in B\} = B$ for all $t \geq 0$. As A attracts B, it follows that $B \subset A$.
The other inclusion is immediate from the defining properties of global attractors. $\square$

For $\phi \in A$, we denote the unique solution $x : \mathbb{R} \to \mathbb{R}$ of equation (1.1) with $x_0 = \phi$ by $x(\phi)$. In case $\phi = 0$ we have $x(\phi)(t) = 0$ for all t. In case $\phi \neq 0$, the solution $x(\phi)$ is bounded, and its zeros form a sequence $(z_j)_{-\infty}^J, J \in \mathbb{Z}$ or $J = \infty$, with property (so). In the last case, the functions

$$[-1, 0] \ni t \mapsto e^{\mu t}|x(z_j + 1 + t)| \in \mathbb{R}$$

are increasing for all j; if $x'(z_j) > 0$ and $t = z_j + 1$ then $x_t \in K$.

PROPOSITION 4.2. *A is connected.*

PROOF. (Compare [**6**, Lemma 2.4.1]). Let $\mathrm{conv}(A \cap \overline{K})$ denote the convex hull of the set $A \cap \overline{K}$ in C. The preceding remarks imply that for each $\phi \in A$ there exist $t < 0$ with $x(\phi)_t \in \overline{K}$. It follows that the bounded set

$$B = A \cup \overline{\mathrm{conv}(A \cap \overline{K})} \subset A \cup \overline{K} \subset \overline{S}$$

is arcwise connected. We have

$$A \subset \{F(t, \psi) : \psi \in A\} \subset \{F(t, \psi) : \psi \in B\} \quad \text{for all} \quad t \geq 0.$$

Suppose A is not connected. Then there exist open disjoint subsets U, V of $\overline{S}$ such that $A \subset U \cup V, A \cap U \neq \emptyset \neq A \cap V$. As A attracts B, there exists $t \geq 0$ such that

$$\{F(t, \psi) : \psi \in B\} \subset U \cup V.$$

Hence

$$\{F(t, \psi) : \psi \in B\} \cap U \supset A \cap U \neq \emptyset \neq A \cap V \subset \{F(t, \psi) : \psi \in B\} \cap V,$$

a contradiction to the fact that $\{F(t, \psi) : \psi \in B\}$ is connected. $\square$

PROPOSITION 4.3. *The map* $F_A : \mathbb{R} \times A \ni (t, \phi) \rightarrow x(\phi)_t \in A$ *is a continuous flow.*

PROOF. It is easy to verify the algebraic properties of a flow for F_A. In order to show continuity, note that each map $F(t, \cdot), t \geq 0$, defines a continuous injective map F_t of A onto A. Compactness implies that the inverse maps F_t^{-1} are continuous. Observe that for $t \leq 0$ and $\phi \in A$, $F_{-t}^{-1}(\phi) = x(\phi)_t$. Let $t \in \mathbb{R}, \phi \in A$ be given. Choose $r < \min\{t, 0\}$. For $s > r$ and $\psi \in A$, $F_A(s, \psi)$ can be written as $F \circ G(s, \psi)$ where $G(s, \psi) = (s - r, F_{-r}^{-1}(\psi))$. This implies continuity at (t, ϕ). $\square$

COROLLARY 4.1. *The map* $\mathbb{R} \times A \ni (t, \phi) \mapsto x(\phi)_t' \in C$ *is continuous. For all* $t \in \mathbb{R}$ *and* $\phi \in A \setminus \{0\}$, $x(\phi)_t' \neq 0$.

PROOF. Continuity is obvious from

$$x(\phi)_t' = D_1 F(2, x(\phi)_{t-2})1 = D_1 F(2, F_A(t - 2, \phi))1.$$

For $\phi \neq 0$, $x = x(\phi)$ is slowly oscillating. Suppose $x_t' = 0$. Then $x(t) = x(t - 1)$, and

$$0 = x'(t) = -\mu x(t) + f(x(t - 1)) = -\mu x(t) + f(x(t)).$$

Hence $0 = x(t) = x(t - 1)$, a contradiction to property (so). $\square$

We turn to the map P. Set $K^* = \overline{P(\overline{K})}$. It follows from [**6**, Theorem 2.4.2] that the set

$$A(P) = \cap_0^\infty P^n(K^*)$$

is compact and invariant. In fact, $A(P)$ is the global attractor of P. The arguments from the proof of [**6**, Lemma 2.4.1] show that A is connected.

PROPOSITION 4.4. $A \cap \overline{K} = A(P)$.

PROOF. 1. Let $\phi \in A \cap \overline{K}$. In case $\phi = 0 = P(0) \in P(\overline{K}) \subset K^*$, we have $\phi = P^n(\phi)$ for all $n \in \mathbb{N}_0$, hence $\phi \in A(P)$. In case $\phi \neq 0$, consider $x = x(\phi)$ and its zero sequence $(z_j)_{-\infty}^J$. We may assume $z_0 = -1$ and $J \geq 0$. Then

$$X(z_{-2n} + 1) = P(X(z_{-2n-2} + 1)) \in P(K) \subset K^* \quad \text{for all} \quad n \in \mathbb{N}_0.$$

Consequently,

$$\phi = X(0) = X(z_0 + 1) \in \cap_0^\infty P^n(K^*).$$

2. Let $\phi \in A(P)$. If $\phi = 0$ then $\phi \in A$ since the zero solution has segments in A, according to Proposition 4.1. Assume $\phi \neq 0$. There is a trajectory $(\phi_n)_{-\infty}^0$ of the map P in K with $\phi_0 = \phi$. It follows that there exists a solution $x : \mathbb{R} \to \mathbb{R}$ of equation (1.1) which is slowly oscillating and has a zero sequence $(z_j)_{-\infty}^J$ with $J \geq 0$ and $z_0 = -1$ so that the phase curve X satisfies

$$X(z_{2n} + 1) = \phi_n \quad \text{for all integers} \quad n \leq 0.$$

Proposition 3.2(ii) implies that for all z_j with $x'(z_j) < 0$,

$$|x(t)| \leq (1 + \mu) \sup f \quad \text{for} \quad z_j \leq t \leq z_j + 1.$$

The monotonicity assertion of Proposition 3.2 shows that the same estimate holds on the interval (z_j, z_{j+1}), in case $j < J$, and on the interval (z_j, ∞) in case $j = J$. It follows that $|x|$ is also uniformly bounded on all intervals (z_k, z_{k+1}) with $x'(z_k) > 0$ and $k < J$, and finally that x is globally bounded. Now Proposition 4.1 yields $\phi = x_0 \in A$. $\square$

CHAPTER 5

Phase Space Decomposition

The solutions $x : [-1, \infty) \to \mathbb{R}$ of equation (1.2), $\alpha = -f'(0)$ as in the introduction, define a C_0–semigroup of operators $T(t) = D_2 F(t, 0)$, $t \geq 0$. The spectrum Σ of its generator consists of complex conjugate pairs of eigenvalues in the double strips S_k given by

$$2k\pi < |\text{Im}(\lambda)| < 2k\pi + \pi, \quad k \in \mathbb{N},$$

and by at most two eigenvalues in the strip S_0 given by

$$|\text{Im}(\lambda)| < \pi;$$

the total multiplicity of Σ in S_0 is 2.

We have

$$\max \Re(\cup_\mathbb{N}(\Sigma \cap S_k)) < \min \Re(\Sigma \cap S_0).$$

Let L and Q denote the reellified generalized eigenspaces associated with the spectral sets $\Sigma \cap S_0$ and $\cup_\mathbb{N}(\Sigma \cap S_k)$, respectively. Then

$$(1.3) \qquad\qquad C = L \oplus Q,$$

$\dim L = 2$, and both L and Q are positively invariant under the maps $T(t)$.

Three cases are possible for $\Sigma \cap S_0$:

I. $\alpha e^\mu < \frac{1}{e}$. $\Sigma \cap S_0$ consists of two simple, real eigenvalues $u_{00} < u_0 < 0$; a basis $\{\beta_1, \beta_2\}$ of L is given by the restrictions of the functions

$$t \mapsto e^{u_0 t}, \quad t \mapsto e^{u_{00} t}$$

to the interval $[-1, 0]$.

II. $\alpha e^\mu = \frac{1}{e}$. $\Sigma \cap S_0$ consists of a double eigenvalue

$$u_0 = -1 - \mu;$$

a basis $\{\beta_1, \beta_2\}$ of L is given by

$$t \mapsto e^{u_0 t}, \quad t \mapsto -t e^{u_0 t}.$$

III. $\alpha e^{\mu} > \frac{1}{e}$. $\Sigma \cap S_0$ consists of a complex conjugate pair $\lambda_0 = u_0 + iv_0$, $\overline{\lambda_0}$ of simple eigenvalues; $0 < v_0$. A basis $\{\beta_1, \beta_2\}$ of L is given by

$$t \mapsto e^{u_0 t} \sin(v_0 t), \quad t \mapsto e^{u_0 t} \cos(v_0 t).$$

For later use we define I to be the isomorphism of L onto the $\mathbb{R}$–vectorspace $\mathbb{C}$ which maps β_1 onto 1, and β_2 onto i.

REMARK 5.1.   (i) *Slowly and rapidly oscillating solutions. The space L consists of the segments x_t of the linear combinations of the functions $\mathbb{R} \to \mathbb{R}$ used to define the bases above. All of these linear combinations (except the trivial one) are slowly oscillating solutions of equation (1.2); we have*

$$L \subset S \cup \{0\}.$$

Real–valued solutions associated with eigenvalues $\lambda = u + iv$ outside S_0 have zeros spaced at distances

$$\frac{\pi}{v} < \frac{1}{2};$$

hence they are not slowly oscillating.
(ii) *Stability. Consider*

$$v(\mu) \in \left(\frac{\pi}{2}, \pi\right) \quad \text{defined by} \quad v(\mu) = -\mu \tan(v(\mu)).$$

For

$$\alpha < -\frac{\mu}{\cos(v(\mu))} \quad \left(= \frac{v(\mu}{\sin(v(\mu))} > \frac{\pi}{2}\right),$$

we have

$$\Re \lambda < 0 \quad \text{for all} \quad \lambda \in \Sigma;$$

and the zero solution of equation (1.2) is exponentially attractive. It follows that there exists a neighbourhood U of 0 and a positive constant c such that

$$\|F(t, \phi)\| \leq c e^{\frac{u_0 t}{2}} \quad \text{for all} \quad \phi \in U, t \geq 0.$$

At $\alpha = -\frac{\mu}{\cos(v(\mu))}$, $u_0 = 0$, and for $\alpha > -\frac{\mu}{\cos(v(\mu))}$, $u_0 > 0$.

Let $p : C \to C$ denote the projection onto L given by (1.3).

LEMMA 5.1. $0 \notin pS$.

PROOF. In case III, see the proof of [**19**, Lemma 6.3]. Analogous arguments work in the cases I and II where nontrivial solutions with phase curve in L have at most one zero. $\square$

In other words,
$$S \cap Q = \emptyset,$$
the wedge S contains the set $L \setminus \{0\}$ and stays away from the complementary subspace Q.

A–Priori Estimates, Phase Curves with Trivial α–Limit Set, and Invariant Manifolds

[**20**, Proposition 10.1], or [**19**, Section 7], yield the following a–priori estimate.

PROPOSITION 6.1. *There is a constant $c_A > 0$ with the following property. If $x : [t_0 - 1, \infty) \to \mathbb{R}$ and $y : [t_0 - 1, \infty) \to \mathbb{R}$ are solutions of eq. (1.1) which are bounded by $\max_A \|\phi\|$, and if $d = x - y$ has no zero in the interval $[t_0 - 1, t_0]$ then, for all $t \geq t_0 + 2$,*

$$\|d_t\| \leq c_A \|pd_t\|.$$

COROLLARY 6.1. *Let $x : \mathbb{R} \to \mathbb{R}$ be a solution of equation (1.1) with phase curve in A. Then*

$$\|x_t\| \leq c_A \|px_t\| \quad \text{for all} \quad t \in \mathbb{R}.$$

PROOF. Either $x(t) = 0$ for all t, or x is slowly oscillating. Consider the second case. Let $t \in \mathbb{R}$. There exists $t_0 \leq t - 2$ so that x has no zero in the interval $[t_0 - 1, t_0]$. Set $y(t) = 0$ for $t \geq t_0 - 1$. Apply Proposition 6.1. □

Corollary 6.1 is needed for the subsequent results about solutions x with $\alpha(x) = \{0\}$, which will later be used in the proof that A is a graph. Proposition 6.1 will imply that A is Lipschitz continuous.

PROPOSITION 6.2. *Phase curves of slowly oscillating solutions $x : \mathbb{R} \to \mathbb{R}$ of equation (1.1) with $\alpha(x) = \{0\}$ have their values in A.*

PROOF. A slowly oscillating solution $x : \mathbb{R} \to \mathbb{R}$ with $\alpha(x) = \{0\}$ is bounded on the interval $(-\infty, 0]$. Proposition 3.1(ii) gives $\inf x^{-1}(0) = -\infty$. It follows that the zeros of x form a sequence $(z_j)_{-\infty}^{J}, J \in \mathbb{Z}$ or $J = \infty$, with property (so). Arguments as in part 2 of the proof of Proposition 4.4 imply that x is bounded. Apply Proposition 4.1. □

PROPOSITION 6.3. *Let $u_0 > 0$. There is a 2–dimensional C^1–submanifold W_0 of C with $T_0 W_0 = L$ such that for every slowly oscillating solution $x : \mathbb{R} \to \mathbb{R}$ of equation (1.1) with $\alpha(x) = \{0\}$ there exists $t \in \mathbb{R}$ with $x_s \in W_0$ for all $s \leq t$.*

PROOF. Recall from [**19**] that there is a 2–dimensional C^1–submanifold W_0 of C which is locally positively invariant for the map $F(1, \cdot)$, and satisfies $T_0 W_0 = L$. For a constant $\beta > 1$ such that

$$e^{u_1} < \beta < e^{u_0}$$

and for an open neighbourhood U of 0 in C we have

$$
\begin{aligned}
W_0 \ = \ & \{\phi \in U : \text{There is a trajectory } (\phi_n)^0_{-\infty} \text{ of } F(1, \cdot) \\
& \text{in } U \text{ such that } \phi_0 = \phi, \ \beta^{-n}\phi_n \in U \text{ for all } n \in -\mathbb{N}_0, \\
& \beta^{-n}\phi_n \to 0 \text{ as } n \to -\infty\}.
\end{aligned}
$$

Recall also that the maps $T(t), t \geq 0$, induce a group of isomorphisms $T_L(t) \in L_c(L, L), t \geq 0$, and that there exists an equivalent norm $|\cdot|$ on C so that the isomorphism $T_L(1)$ induced by $T(1) - D_2 F(1, 0)$ satisfies

$$|T_L(1)^{-1}| = \sup\{|T_L(1)^{-1}\phi| : |\phi| = 1\} < \beta^{-1}.$$

Consider a slowly oscillating solution $x : \mathbb{R} \to \mathbb{R}$ with $\alpha(x) = \{0\}$. Proposition 6.2 says that $x_t \in A$ for all t. Choose $r > 0$ with $\{\phi \in C : |\phi| \leq r\} \subset U$. Set

$$\hat{\beta} = \frac{1}{2}(|T_L(1)^{-1}|^{-1} + \beta) > \beta.$$

Convergence of x_t to 0 as $t \to -\infty$ and differentiability imply that there exists $t \in \mathbb{R}$ such that for all $s \leq t$, we have $c_A|px_s| < r, x_s \in U$, and

$$|p(F(1, x_s) - D_2 F(1, 0)x_s)| \leq \frac{|T_L(1)^{-1}|^{-1} - \beta}{2c_A}|x_s|.$$

For such s,

$$
\begin{aligned}
|px_{s+1}| = |pF(1, x_s)| \ & \geq \ |pD_2 F(1, 0)x_s| - |p(\dots)| \\
& = \ |D_2 F(1, 0)px_s| - |p(\dots)| \\
& = \ |T_L(1)px_s| - |p(\dots)| \\
& \geq \ |T_L(1)^{-1}|^{-1}|px_s| - \frac{|T_L(1)^{-1}|^{-1} - \beta}{2c_A}|x_s| \\
& \geq \ (|T_L(1)^{-1}|^{-1} - \frac{1}{2}(|T_L(1)^{-1}|^{-1} - \beta))|px_s| \\
& \qquad \text{(see Corollary 6.1)} \\
& = \ \hat{\beta}|px_s|.
\end{aligned}
$$

Let $s \leq t$. It follows that for all $n \in \mathbb{N}_0$, we have $|px_{s-n}| \leq \hat{\beta}^{-n}|px_s|$. Using Corollary 6.1 again we infer

$$\beta^n|x_{s-n}| \leq c_A\beta^n|px_{s-n}| \leq c_A\beta^n\hat{\beta}^{-n}|px_s| \quad \text{for all} \quad n \in \mathbb{N}_0.$$

In particular, $|\beta^n x_{s-n}| < r$ for all $n \in \mathbb{N}_0$, and $\beta^n x_{s-n} \to 0$ as $n \to \infty$. Set $\phi_n = x_{s+n}$, for $n \in -\mathbb{N}_0$. According to the preceding characterization of W_0, we get $x_s = \phi_0 \in W_0$. $\quad\square$

We also need the analogue of Proposition 6.3 in the case $u_0 = 0$. This requires center manifolds. We briefly review the construction of center manifolds for functional differential equations in [**2**]. First we have to recall a few basic facts about dual semigroups.

The elements $\phi^\odot \in C^*$ for which the adjoints of the operators $T(t)$ define a continuous curve

$$\mathbb{R}^+ \ni t \mapsto T(t)^* \phi^\odot \in C^*$$

form a positively invariant subspace $C^\odot$, which is called the *sun subspace*. The operators

$$C^\odot \ni \phi^\odot \mapsto T(t)^* \phi^\odot \in C^\odot, \quad t \geq 0,$$

constitute a C_0–semigroup $T^\odot$ on $C^\odot$. Using $T^\odot$ we define the space $C^{\odot\odot} \subset C^{\odot*}$. The phase space C is sun–reflexive with respect to the semigroup T in the sense that there exists a norm–preserving isomorphism of C onto $C^{\odot\odot}$.

There is an isomorphism between $C^{\odot*}$ and $\mathbb{R} \times L^\infty(-1, 0; \mathbb{R})$. Let $r^{\odot*} \in C^{\odot*}$ denote the element which corresponds to $(1, 0) \in \mathbb{R} \times L^\infty(-1, 0; \mathbb{R})$.

For a given continuous function $\tilde{g} : \mathbb{R} \to C^{\odot*}$ and reals $a \leq b$ the weak–star integral

$$\int_a^b T^\odot(b-t)^* \tilde{g}(t) dt \in C^{\odot*}$$

is defined by

$$\left(\int_a^b T^\odot(b-t)^* \tilde{g}(t) dt \right)(x^\odot) = \int_a^b (T^\odot(b-t)^* \tilde{g}(t))(x^\odot) dt$$

for $x^\odot \in C^\odot$. If $g : \mathbb{R} \to \mathbb{R}$ is a continuous function and if $x : \mathbb{R} \to \mathbb{R}$ is a solution of the inhomogeneous equation

(6.1) $$x'(t) = -\mu x(t) - \alpha x(t-1) + g(t)$$

(where $\alpha = -f'(0)$), then the curve $u : \mathbb{R} \ni t \mapsto x_t \in C$ is a solution of the integral equations

(6.2) $$u(t) = T(t-s)u(s) + \int_s^t T^\odot(t-\tau)^* (g(\tau) r^{\odot*}) d\tau, \quad t \geq s.$$

The last equation is in fact an equation between elements of $C^{\odot*}$; the inclusion map $C \cong C^{\odot\odot} \to C^{\odot*}$ has been omitted.

Next, we prepare modified nonlinearities for the construction of a center manifold. There are sequences of C^1–functions $r_n : \mathbb{R} \to \mathbb{R}$ with compact supports and of open intervals $I_n, n \in \mathbb{N}$, such that for all $n \in \mathbb{N}$, $0 \in I_n$ and

$$f(\xi) = f'(0)\xi + r_n(\xi) \quad \text{for} \quad \xi \in I_n,$$

and

$$|r_n(\xi) - r_n(\xi')| \leq \frac{1}{n} |\xi - \xi'| \quad \text{for} \quad \xi, \xi' \quad \text{in} \quad \mathbb{R}.$$

For a given real Banach space E and $\eta > 0$, let $BC^\eta(\mathbb{R}, E)$ denote the Banach space of continuous maps $u : \mathbb{R} \to E$ such that

$$\|u\|_\eta = \sup\{e^{-\eta|t|}\|u(t)\| : t \in \mathbb{R}\} < \infty.$$

The equations $R_n(u)(t) = r_n(u(t)(-1))r^{\odot*}$ define substitution operators

$$R_n : BC^\eta(\mathbb{R}, C) \to BC^\eta(\mathbb{R}, C^{\odot*}), \quad n \in \mathbb{N},$$

which are Lipschitz continuous. There are Lipschitz constants L_n of $R_n, n \in \mathbb{N}$, so that

$$L_n \to 0 \quad \text{as} \quad n \to \infty.$$

Assume $u_0 = 0$ from now on. This implies that there exists a constant $M \geq 1$ with

(6.3) $$\|T(t)\| \leq M \quad \text{for} \quad t \geq 0.$$

Fix $\eta > 0$ with $u_1 < -\eta < 0$. For each $\hat{F} \in BC^\eta(\mathbb{R}, C^{\odot*})$ there exists a unique solution

$$u = \hat{K}(\hat{F}) \in BC^\eta(\mathbb{R}, C)$$

of the integral equations

(6.4) $$u(t) = T(t - s)u(s) + \int_s^t T^\odot(t - \tau)^* \hat{F}(\tau)d\tau, \quad t \geq s,$$

with the property

$$pu(0) = 0.$$

The solution map $\hat{K} : BC^\eta(\mathbb{R}, C^{\odot*}) \to BC^\eta(\mathbb{R}, C)$ is linear and continuous.

Fix an integer n so large that

$$\frac{M}{n}\|r^{\odot*}\| < \eta, \quad L_n\|\hat{K}\| < \frac{1}{2}.$$

For every $\phi \in L$ there exists a unique solution $u = u(\phi) \in BC^\eta(\mathbb{R}, C)$ of the equation

$$u = T_L(\cdot)\phi + \hat{K}(R_n(u)).$$

The points $u(\phi)(0) \in C$ form a 2–dimensional graph W_n^c of class C^1 in $C = L \oplus Q$ which satisfies

$$T_0 W_n^c = L.$$

The graph W_n^c is called the global center manifold of the modified equation

(6.5) $$y'(t) = -\mu y(t) - \alpha y(t - 1) + r_n(y(t - 1)).$$

PROPOSITION 6.4. *Let $u_0 = 0$. There is a 2–dimensional C^1–submanifold W^c of C with $T_0 W^c = L$ so that for every solution $x : \mathbb{R} \to \mathbb{R}$ of equation (1.1) with $\alpha(x) = \{0\}$ there are $t_0 \in \mathbb{R}$ such that for all $t \leq t_0$, $x_t \in W^c$.*

PROOF. 1. Consider a graph $W^c = W_n^c$ as above. Let a solution $x : \mathbb{R} \to \mathbb{R}$ of equation (1.1) be given so that $\alpha(x) = \{0\}$. Choose $t_0 \in \mathbb{R}$ such that for $t \le t_0$, $x(t) \in I_n$. Consider some $t \le t_0$. There exists a solution $\tilde{x} : \mathbb{R} \to \mathbb{R}$ of the modified equation (6.5) which coincides with x on the interval $(-\infty, t]$. For $s \in \mathbb{R}$, set $u(s) = \tilde{x}_{s+t} \in C$. The map $u : \mathbb{R} \to C$ satisfies

$$(6.6) \qquad u(s) = T(s - s')u(s') + \int_{s'}^{s} T^{\odot}(s - \sigma)^*(r_n(u(\sigma)(-1))r^{\odot *})d\sigma$$

for $s \ge s'$, according to the remarks above concerning the equations (6.1) and (6.2).

2. Proof of $u \in BC^{\eta}(\mathbb{R}, C)$. The map u is bounded on the interval $(-\infty, 0]$ since $x(s) \to 0$ as $s \to -\infty$. Therefore

$$\sup\{e^{-\eta|s|}\|u(s)\| : s \le 0\} < \infty.$$

Let $s \ge 0$. Then the relations

$$\|T^{\odot *}(\sigma)\| = \|T^{\odot}(\sigma)\| \le \|T^*(\sigma)\| = \|T(\sigma)\| \le M \quad \text{for} \quad \sigma \ge 0,$$

the definition of the weak–star integral, and equation (6.6) altogether imply

$$\|u(s)\| \le M\|u(0)\| + \int_0^s \frac{M}{n}|u(\sigma)(-1)|\|r^{\odot *}\|d\sigma.$$

Using Gronwall's lemma and the choice of n, we get

$$\|u(s)\| \le M\|u(0)\|e^{\frac{M}{n}\|r^{\odot *}\|s} \le M\|u(0)\|e^{\eta s} \quad \text{for} \quad s \ge 0,$$

hence

$$\sup\{e^{-\eta|s|}\|u(s)\| : s \ge 0\} < \infty.$$

3. Now equation (6.6) can be rewritten as

$$(6.7) \quad u(s) = T(s - s')u(s') + \int_{s'}^{s} T^{\odot}(s - \sigma)^* R_n(u)(\sigma)d\sigma \quad \text{for} \quad s \ge s'.$$

The map $v : \mathbb{R} \ni s \mapsto u(s) - T_L(s)pu(0) \in C$ is in $BC^{\eta}(\mathbb{R}, C)$ since $s \mapsto T_L(s)pu(0)$ is bounded. We have $pv(0) = 0$, and v is a solution of equation (6.7), i.e.,

$$v(s) = T(s - s')v(s') + \int_{s'}^{s} T^{\odot}(s - \sigma)^* R_n(u)(\sigma)d\sigma \quad \text{for} \quad s \ge s'.$$

The remarks about solutions of equation (6.4) imply

$$v = \hat{K}(R_n(u)).$$

Therefore

$$u = T_L(\cdot)pu(0) + v = T_L(\cdot)pu(0) + \hat{K}(R_n(u)),$$

and we conclude that

$$x_t = \tilde{x}_t = u(0) \in W^c.$$

$\square$

CHAPTER 7

Graph Representation

Now we can state the main result.

THEOREM 7.1.   (i) *There exists a Lipschitz continuous map* $a : pA \to Q$ *such that*

$$(7.1) \qquad\qquad A = \{\chi + a(\chi) : \chi \in pA\}.$$

(ii) *In case* $A \neq \{0\}$ *there exists a slowly oscillating periodic solution* $y :$ $\mathbb{R} \to \mathbb{R}$ *of equation* (1.1), *with minimal period* $\tau > 0$, *such that the simple closed curve* $\eta : [0, \tau] \ni t \mapsto y_t \in C$ *with trace in* A *satisfies*

$$pA = \overline{\mathrm{int}(p \circ \eta)}.$$

In this section we prove Theorem 7.1(i). The existence of a map $a : pA \to Q$ such that (7.1) holds is equivalent to injectivity of the restriction $p|A$. In order to show injectivity of $p|A$ it is sufficient to prove that for all ϕ, ψ in A with $\phi \neq \psi$ we have

$$(7.2) \qquad\qquad \phi - \psi \in S$$

since (7.2) implies

$$\begin{aligned} 0 \;&\neq\; p(\phi - \psi) \quad \text{(see Lemma 5.1)} \\ &=\; p\phi - p\psi. \end{aligned}$$

A relatively simple case where (7.2) holds is covered by [**20**, Proposition 6.1]. We restate this as

REMARK 7.1. *If* x *and* y *are slowly oscillating periodic solutions of equation* (1.1), *and if* $x_t - y_s \neq 0$ *for some* t, s *in* $\mathbb{R}$, *then*

$$x_t - y_s \in S.$$

PROOF OF THE FIRST ASSERTION OF THEOREM 7.1. 1. Existence of a map $a : pA \to Q$ so that (7.1) holds. Let ϕ, ψ in A be given, with $\phi \neq \psi$. According to the discussion preceding Remark 7.1 it suffices to show that (7.2) holds. For $t \in \mathbb{R}$, set $X(t) = x(\phi)_t, Y(t) = x(\psi)_t$. Recall that $X(t) \neq Y(t)$ for all $t \in \mathbb{R}$.

1.1. The case $\alpha(x(\phi)) = \{0\} = \alpha(x(\psi))$. Either $\phi \neq 0$, or $\psi \neq 0$. In each case there is a nonzero solution of equation (1.1) which tends to 0 as $t \to -\infty$. Using Remark 5.1(ii) we infer $u_0 \geq 0$.

1.1.1. The case $u_0 > 0$. Proposition 6.3 implies that for some $t \geq 2$, both $X(-t)$ and $Y(-t)$ belong to the set $U_0 \cap W_0$ where U_0 is the neighbourhood of 0 in C obtained in [**19**, Proposition 6.4]; therefore

$$X(-t + s) - Y(-t + s) \in S \quad \text{for some} \quad s \in [0, 2].$$

Proposition 3.3 gives

$$\phi - \psi = X(0) - Y(0) \in S.$$

1.1.2. The case $u_0 = 0$. The proof of [**19**, Proposition 6.4] is valid also for the manifold W^c. It follows that there is a neighbourhood U^c of 0 in C such that for every pair of different points χ, ρ in $U^c \cap W^c$ there exists $s \in [0, 2]$ with

$$F(s, \chi) - F(s, \rho) \in S.$$

Proposition 6.4 (of the present paper) implies that for some $t \geq 2$, the different points $X(-t), Y(-t)$ belong to $U^c \cap W^c$. Hence $X(s - t) - Y(s - t) \in S$ for some $s \in [0, 2]$, and by Proposition 3.3,

$$\phi - \psi = X(0) - Y(0) \in S.$$

1.2. The case $\alpha(x(\phi)) \neq \{0\}$. The compactness of A implies that there exist $\chi \neq 0$ and ρ in A, and a sequence $t_n \to -\infty$ such that

$$X(t_n) \to \chi, \quad Y(t_n) \to \rho \quad \text{as} \quad n \to \infty.$$

For $t \in \mathbb{R}$, set $W(t) = x(\chi)_t$ and $Z(t) = x(\rho)_t$.

1.2.1. The case $\rho = 0$. We have $0 \neq \chi \in A \subset \overline{S} = S \cup \{0\}$. Hence $\chi \in S$. By Proposition 3.3, there exists $t \in [0, 4]$ so that the function $W(t) = W(t) - Z(t)$ has no zero in its domain $[-1, 0]$. For n sufficiently large, $t_n + t < 0$, and the function

$$X(t_n + t) - Y(t_n + t) = F_A(t, X(t_n)) - F_A(t, Y(t_n))$$

$$\approx F_A(t, \chi) - F_A(t, \rho) = W(t) - Z(t)$$

has no zero. Therefore,

$$X(t_n + t) - Y(t_n + t) \in S,$$

and by Proposition 3.3,

$$\phi - \psi = X(0) - Y(0) \in S.$$

1.2.2. The case $\rho \neq 0$ and $\chi - \rho \neq 0$. In order to obtain (7.2) it is sufficient to derive

(7.3) $$\chi - \rho \in \overline{S}$$

since (7.3) and $\chi - \rho \neq 0$ imply $\chi - \rho \in S$, and we can argue as in case 1.2.1: For some $t \in [0, 4]$, $F(t, \chi) - F(t, \rho)$ has no zero (Proposition 3.3); for n sufficiently large, $t + t_n < 0$, and $X(t + t_n) - Y(t + t_n)$ has no zero; finally

$$\phi - \psi = X(0) - Y(0) \in S \quad \text{(Proposition 3.3)}.$$

In order to show (7.3) it is sufficient to prove that

(7.4) $$\text{there exists} \quad t < 0 \quad \text{with} \quad W(t) - Z(t) \in \overline{S}$$

since this implies

$$\chi - \rho = F(-t, W(t)) - F(-t, Z(t)) \in \overline{S},$$

by Proposition 3.3 once again.

Proof of (7.4). There exist $t < 0$ so that the function $W(t) : [-1, 0] \to \mathbb{R}$ is strictly positive, and $s < t$ so that $Z(s)$ is strictly negative. Hence, $W(t) - Z(s)$ is strictly positive. It follows that there exists $\epsilon' > 0$ such that for $0 \leq t' \leq \epsilon'$,

$$W(t) - Z(s + t') \quad \text{is strictly positive.}$$

In particular,

$$W(t) - Z(s + t') \in S.$$

Suppose $W(t) - Z(t) \notin \overline{S}$. We derive a contradiction. Note first that the assumption implies that there exist $\epsilon \geq \epsilon'$ and a sequence $\delta_n \searrow 0$ such that

$$s + \epsilon < t,$$

$$W(t) - Z(s + t') \in \overline{S} \quad \text{for} \quad 0 \leq t' \leq \epsilon,$$

(7.5) $$W(t) - Z(s + \epsilon + \delta_n) \notin \overline{S} \quad \text{for} \quad n \in \mathbb{N}.$$

1.2.2.1. The case $W(t) - Z(s + \epsilon) \neq 0$. Then $W(t) - Z(s + \epsilon) \in S$. By Proposition 3.3, there exists $t_\epsilon \in [0, 4]$ so that $W(t + t_\epsilon) - Z(s + \epsilon + t_\epsilon)$ has no zero. Choose $r > 0$ such that for all $\overline{\phi}, \overline{\psi}$ in C with

$$\|\overline{\phi} - W(t + t_\epsilon)\| < r, \quad \|\overline{\psi} - Z(s + \epsilon + t_\epsilon)\| < r$$

the function $\overline{\phi} - \overline{\psi}$ has no zero, and therefore belongs to the set S. Choose $\delta > 0$ such that for $0 \leq \theta \leq \delta$,

$$\|Z(s + \epsilon + t_\epsilon + \theta) - Z(s + \epsilon + t_\epsilon)\| < \frac{r}{2}.$$

Recall from Proposition 4.3 that for $n \to \infty$,

$$X(t + t_\epsilon + t_n) = F_A(t + t_\epsilon, X(t_n)) \to F_A(t + t_\epsilon, \chi) = W(t + t_\epsilon),$$

and
$$Y(s + \epsilon + t_\epsilon + t_n) = F_A(s + \epsilon + t_\epsilon, Y(t_n)) \to F_A(s + \epsilon + t_\epsilon, \rho) = Z(s + \epsilon + t_\epsilon);$$

moreover, by continuous dependence of solutions on initial data,
$$Y(s + \epsilon + t_\epsilon + \theta + t_n) \to Z(s + \epsilon + t_\epsilon + \theta) \quad \text{as} \quad n \to \infty,$$
$$\text{uniformly for} \quad 0 \le \theta \le \delta.$$

It follows that there exists an integer n_0 such that for all $n \ge n_0$ in $\mathbb{N}_0$ and for all $\theta \in [0, \delta]$,
$$\|X(t + t_\epsilon + t_n) - W(t + t_\epsilon)\| < r$$

and
$$\|Y(s + \epsilon + t_\epsilon + \theta + t_n) - Z(s + \epsilon + t_\epsilon)\| \le$$
$$\|Y(s + \epsilon + t_\epsilon + \theta + t_n) - Z(s + \epsilon + t_\epsilon + \theta)\| + \|Z(s + \epsilon + t_\epsilon + \theta) - Z(s + \epsilon + t_\epsilon)\| < r.$$

Consequently, for such n and θ,
$$X(t + t_\epsilon + t_n) - Y(s + \epsilon + t_\epsilon + \theta + t_n) \in S.$$

Let $\theta \in [0, \delta]$ be given. Consider an integer $n \ge n_o$. Choose an integer $k \ge n$ so large that
$$t_n - t_k - t_\epsilon \ge 0.$$

We infer, using Proposition 3.3, that
$$S \ni X(t + t_\epsilon + t_k + (t_n - t_k - t_\epsilon)) - Y(s + \epsilon + t_\epsilon + \theta + t_k + (t_n - t_k - t_\epsilon))$$
$$= X(t + t_n) - Y(s + \epsilon + \theta + t_n).$$

It follows that
$$\overline{S} \ni \lim_{n \to \infty} F_A(t, X(t_n)) - \lim_{n \to \infty} F_A(s + \epsilon + \theta, Y(t_n)) = W(t) - Z(s + \epsilon + \theta)$$

for all $\theta \in [0, \delta]$, which is a contradiction to (7.5).

1.2.2.2. The case $W(t) - Z(s + \epsilon) = 0$. The zeros of $x(\rho)$ are either unbounded on $\mathbb{R}^+$, simple and spaced at distances larger than 1, or they are bounded from above in which case $x(\rho)$ and Z tend to 0 as $t \to \infty$. In both cases there exists $\sigma > t$ so that the function $W(t) - Z(\sigma)$ is strictly positive. Consequently, there exists $\epsilon'' > 0$ so that for $0 \le t' \le \epsilon''$,
$$W(t) - Z(\sigma - t') \quad \text{is strictly positive};$$

in particular,
$$W(t) - Z(\sigma - t') \in S.$$

The assumption $W(t) - Z(t) \notin \overline{S}$ implies that there exist $\epsilon^* \ge \epsilon''$ and a sequence $\delta_n^* \searrow 0$ so that
$$\sigma - \epsilon^* > t,$$
$$W(t) - Z(\sigma - t') \in \overline{S} \quad \text{for} \quad 0 \le t' \le \epsilon^*,$$

(7.6) $$W(t) - Z(\sigma - \epsilon^* - \delta_n^*) \notin \overline{S} \quad \text{for} \quad n \in \mathbb{N}.$$

1.2.2.2.1. The case $W(t) - Z(\sigma - \epsilon^*) = 0$. Then $Z(\sigma - \epsilon^*) = W(t) = Z(s + \epsilon)$, and $s + \epsilon < t < \sigma - \epsilon^*$. It follows that $x(\rho)$ and Z are periodic, and W is a translate of Z. Remark 7.1 gives

$$W(t) - Z(t) \in \overline{S},$$

a contradiction to the assumption (7.5).

1.2.2.2.2. The case $W(t) - Z(\sigma - \epsilon^*) \neq 0$. Then $W(t) - Z(\sigma - \epsilon^*) \in S$, and we can proceed as in case 1.2.2.1. The details are as follows. There exists $t^* \in [0, 4]$ so that $W(t + t^*) - Z(\sigma - \epsilon^* + t^*)$ has no zero. There exists $r^* > 0$ such that for $\overline{\phi}, \overline{\psi}$ in C with

$$\|\overline{\phi} - W(t + t^*)\| < r^*, \quad \|\overline{\psi} - Z(\sigma - \epsilon^* + t^*)\| < r^*,$$

the function $\overline{\phi} - \overline{\psi}$ has no zero, and therefore belongs to S. Proposition 4.3 yields a constant $\delta^* > 0$ so that for $0 \leq \theta \leq \delta^*$,

$$\|Z(\sigma - \epsilon^* + t^* - \theta) - Z(\sigma - \epsilon^* + t^*)\| < \frac{r^*}{2}.$$

We have, for $n \to \infty$,

$$X(t + t^* + t_n) = F_A(t + t^*, X(t_n)) \to F_A(t + t^*, \chi) = W(t + t^*),$$

and, using the continuous flow of Proposition 4.3 again,

$$Y(\sigma - \epsilon^* + t^* - \theta + t_n) = F_A(\sigma - \epsilon^* + t^* - \theta, Y(t_n))$$

$$\to F_A(\sigma - \epsilon^* + t^* - \theta, \rho) = Z(\sigma - \epsilon^* + t^* - \theta)$$

$$\text{uniformly for} \quad \theta \in [0, \delta^*].$$

It follows that there exists an integer n_0 such that for $n \geq n_0$ and $0 \leq \theta \leq \delta^*$,

$$\|X(t + t^* + t_n) - W(t + t^*)\| < r^*$$

and

$$\|Y(\sigma - \epsilon^* + t^* - \theta + t_n) - Z(\sigma - \epsilon^* + t^*)\|$$

$$\leq \|Y(\sigma - \epsilon^* + t^* - \theta + t_n) - Z(\sigma - \epsilon^* + t^* - \theta)\|$$

$$+ \|Z(\sigma - \epsilon^* + t^* - \theta) - Z(\sigma - \epsilon^* + t^*)\|$$

$$< \frac{r^*}{2} + \frac{r^*}{2} = r^*.$$

We infer that for such n and θ,

$$X(t + t^* + t_n) - Y(\sigma - \epsilon^* + t^* - \theta + t_n) \in S.$$

Let $\theta \in [0, \delta^*]$ be given. Consider an integer $n \geq n_0$. Choose an integer $k \geq n$ so large that

$$t_n - t_k - t^* \geq 0.$$

Proposition 3.3 and $k \geq n_0$ yield

$$S \ni X(t + t^* + t_k + (t_n - t_k - t^*)) - Y(\sigma - \epsilon^* + t^* - \theta + t_k + (t_n - t_k - t^*))$$

$$= X(t + t_n) - Y(\sigma - \epsilon^* - \theta + t_n).$$

It follows that

$$\overline{S} \ni \lim_{n \to \infty} F_A(t, X(t_n)) - \lim_{n \to \infty} F_A(\sigma - \epsilon^* - \theta, Y(t_n))$$

$$= W(t) - Z(\sigma - \epsilon^* - \theta) \quad \text{for all} \quad \theta \in [0, \delta^*],$$

a contradiction to (7.6).

1.2.3. The case $\rho \neq 0$ and $\chi - \rho = 0$. Since $x(\chi)$ is slowly oscillating there exists $\epsilon_0 > 0$ such that for $0 < \epsilon < \epsilon_0$, we have $x(\chi)'(\epsilon) \neq 0$. It follows that for such ϵ,

$$W(\epsilon) - W(0) = x(\chi)_\epsilon - x(\chi)_0 \neq 0.$$

Let $\epsilon \in (0, \epsilon_0)$. Consider the solutions $x(\phi, \epsilon) : \mathbb{R} \ni t \mapsto x(\phi)(\epsilon + t) \in \mathbb{R}$ and $x(\psi)$ of equation (1.1). We have $x(\phi, \epsilon)_0 = X(\epsilon)$. Therefore

$$F_A(t_n, x(\phi, \epsilon)_0) = F_A(t_n, X(\epsilon)) = X(\epsilon + t_n) = F_A(\epsilon, X(t_n))$$

$$\to F_A(\epsilon, \chi) = W(\epsilon) \quad \text{as} \quad n \to \infty.$$

Recall

$$F_A(t_n, \psi) \to \rho = Z(0) \quad \text{as} \quad n \to \infty.$$

Observe that

$$W(\epsilon) \neq 0, \quad Z(0) = \rho \neq 0, \quad W(\epsilon) - Z(0) \neq 0;$$

i.e., the solutions $x(\phi, \epsilon)$ and $x(\psi)$, the sequence $t_n \to -\infty, W(\epsilon)$, and $Z(0)$ satisfy the hypotheses of subcase 1.2.2. Consequently,

$$x(\phi)_\epsilon - x(\psi)_0 = x(\phi, \epsilon)_0 - \psi \in S.$$

Taking the limit as $\epsilon \searrow 0$ we get

$$(0 \neq \quad) \quad \phi - \psi = x(\phi)_0 - x(\psi)_0 \in \overline{S} = S \cup \{0\}$$

which yields (7.2).

1.3. We have shown that (7.2) holds for every pair of different points in A. It follows that the projection p is injective on the set A, and that there is a map $a : pA \to Q$ such that (7.1) holds.

2. Proof that the map a satisfies a Lipschitz condition. Consider different points χ, ρ in pA. For $t \in \mathbb{R}$, set $X(t) = x(\chi + a(\chi))_t$ and $Y(t) = x(\rho + a(\rho))_t$. We have $X(0) \neq Y(0)$. It follows that $X(-6) \neq Y(-6)$. Both points belong to A. By (7.2), $X(-6) - Y(-6) \in S$. Proposition 3.3 implies that for some $s \in [0, 4]$, $X(s - 6) - Y(s - 6)$ has no zero. Now Proposition 6.1 gives

$$\|X(0) - Y(0)\| \leq c_A \|p(X(0) - Y(0))\|.$$

Finally,

$$
\begin{aligned}
\|a(\chi) - a(\rho)\| &= \|(id - p)(X(0) - Y(0))\| \\
&\leq (1 + \|p\|)\|X(0) - Y(0)\| \\
&\leq (1 + \|p\|)c_A\|pX(0) - pY(0)\| \\
&= (1 + \|p\|)c_A\|\chi - \rho\|.
\end{aligned}
$$

$\square$

CHAPTER 8

Transversals

In this section we use the flow F_A in order to obtain homeomorphisms from subsets in pA of transversals to projected phase curves onto open subsets of the compact space $A(P) \subset \overline{K}$, and homeomorphisms between subsets in pA on different transversals.

PROPOSITION 8.1. *For $t \in \mathbb{R}, 0 \neq \phi \in A$, and $x = x(\phi)$, we have*

$$px_t' = D(s \mapsto px_s)(t)1 \in L \setminus \{0\}.$$

PROOF. According to Corollary 4.1, $x_t' \neq 0$. Part 1.3 of the proof of Theorem 7.1(i) yields

$$\frac{1}{h}(x_{t+h} - x_t) \in \frac{1}{h}(S \cup \{0\}) = \overline{S} \quad \text{for all} \quad h \neq 0.$$

Hence $x_t' \in \overline{S} \setminus \{0\} = S$, and by Lemma 5.1, $0 \neq px_t' = D(s \mapsto px_s)(t)1$. $\square$

PROPOSITION 8.2. *Let $\rho \in pA \setminus \{0\}$ be given. Let $\phi = \rho + a(\rho), x = x(\phi), t \in \mathbb{R}$.*

(i) *Suppose $x_t \in H$. Then there exist an open neighbourhood N of ρ in L, $\epsilon > 0$, and a continuous map $\sigma : N \cap pA \to (t - \epsilon, t + \epsilon)$ such that for all $\tilde{x} = x(\tilde{\rho} + a(\tilde{\rho}))$ with $\tilde{\rho} \in N \cap pA$,*

$$\tilde{x}_s \in H \quad \text{and} \quad |s - t| < \epsilon \quad \text{if and only if} \quad s = \sigma(\tilde{\rho}).$$

(ii) *Suppose the line $T = px_t + \mathbb{R}\chi$, where $\chi \in L \setminus \{0\}$, is transversal to the projected phase curve $s \mapsto px_s$ at $s = t$, i.e.,*

$$px_t' = D(s \mapsto px_s)(t)1 \notin \mathbb{R}\chi.$$

Then there exist an open neighbourhood U of ϕ in C, $\epsilon > 0$, and a continuous map $\sigma : U \cap A \to (t - \epsilon, t + \epsilon)$ such that for all $\tilde{x} = x(\tilde{\phi})$ with $\tilde{\phi} \in U \cap A$ we have

$$p\tilde{x}_s \in T \quad \text{and} \quad |s - t| < \epsilon \quad \text{if and only if} \quad s = \sigma(\tilde{\phi}).$$

41

PROOF. 1. Proof of (i). Recall that x is slowly oscillating. Therefore $x_t \in H$ implies $x(t-1) = 0 \neq x'(t-1)$. According to Proposition 3.5 there exist an open neighbourhood U of $\psi = x_{t-2}$ in C, $\epsilon > 0$, and a C^1-map $\tilde{\sigma} : U \to (2 - \epsilon, 2 + \epsilon)$ such that for all $\tilde{\psi} \in U$,

$$F(s, \tilde{\psi}) \in H \quad \text{and} \quad |s - 2| < \epsilon \quad \text{if and only if} \quad s = \tilde{\sigma}(\tilde{\psi}).$$

The homeomorphism $F_A(t-2, \cdot)$ maps an open neighbourhood $\tilde{U} \cap A, \tilde{U}$ open in C, of ϕ in A onto $U \cap A$. We have $p(\tilde{U} \cap A) = N \cap pA$, with an open neighbourhood N of ρ in L. For $\tilde{\rho} \in N \cap pA$, set $\sigma(\tilde{\rho}) = \tilde{\sigma}(F_A(t - 2, \tilde{\rho} + a(\tilde{\rho}))) + t - 2$.

2. Proof of (ii). As before, set $\psi = x_{t-2}$. Let $\lambda : L \to \mathbb{R}$ denote a linear map with $\lambda^{-1}(0) = \mathbb{R}\chi$. For $s > 1$ and $\tilde{\psi} \in C$, the relation $pF(s, \tilde{\psi}) \in T$ is equivalent to the equation

$$\lambda(pF(s, \tilde{\psi}) - px_t) = 0.$$

The transversality condition yields

$$D_1((s, \tilde{\psi}) \mapsto \lambda(pF(s, \tilde{\psi}) - px_t))(2, \psi)1 = \lambda(px_t') \neq 0$$

so that the Implicit Function Theorem is applicable. There exist an open neighbourhood $\tilde{U}$ of ψ in C, $\epsilon \in (0, 1)$, and a C^1-map $\tilde{\sigma} : \tilde{U} \to (2 - \epsilon, 2 + \epsilon)$ such that $\tilde{\sigma}(\psi) = 2$, and for all $\tilde{\psi} \in \tilde{U}$,

$$pF(s, \tilde{\psi}) \in T \quad \text{and} \quad |s - 2| < \epsilon \quad \text{if and only if} \quad s = \tilde{\sigma}(\tilde{\psi}).$$

The homeomorphism $F_A(t-2, \cdot)$ maps an open neighbourhood $U \cap A$ of ϕ in A, with U open in C, onto $\tilde{U} \cap A$. For $\tilde{\phi} \in U \cap A$, set $\sigma(\tilde{\phi}) = \tilde{\sigma}(F_A(t-2, \tilde{\phi})) + t - 2$. □

COROLLARY 8.1. Let $\rho \in pA \setminus \{0\}$ be given. Let $\phi = \rho + a(\rho), x = x(\phi)$. Suppose the line $T = \rho + \mathbb{R}\chi, 0 \neq \chi \in L$, is transversal to the projected phase curve $s \mapsto px_s$ at $s = 0$, i.e.,

$$px_0' \notin \mathbb{R}\chi.$$

Let $t \in \mathbb{R}$ be given such that $x_t \in K$. Then there exist open neighbourhoods N of ρ in L and U of x_t in C, $\epsilon > 0$, and a continuous map $\sigma : N \cap pA \to (t - \epsilon, t + \epsilon)$ with $\sigma(\rho) = t$ such that the map

$$N \cap pA \cap T \ni \tilde{\rho} \to F_A(\sigma(\tilde{\rho}), \tilde{\rho} + a(\tilde{\rho})) \in A$$

defines a homeomorphism onto $U \cap A \cap H = U \cap A(P)$.

PROOF. 1. The solution x is slowly oscillating. This implies that $x_t \in K$ belongs to the set $H \cap \{\tilde{\psi} \in C : \tilde{\psi}(0) > 0\}$. Proposition 8.1(i) yields an open neighbourhood N_1 of ρ in L, $\epsilon > 0$, and a continuous map $\sigma : N_1 \cap pA \to (t - \epsilon, t + \epsilon)$ such that for $\tilde{\rho} \in N_1 \cap pA$,

$$F_A(s, \tilde{\rho} + a(\tilde{\rho})) \in H \quad \text{and} \quad |s - t| < \epsilon \quad \text{if and only if} \quad s = \sigma(\tilde{\rho}).$$

An application of Proposition 8.1(ii) to $\psi = x_t, \overline{x} = x(\psi)$, and $\overline{t} = -t$ yields an open neighbourhood U_1 of ψ in C, $\overline{\epsilon} > 0$, and a continuous map $\overline{\sigma} : U_1 \cap A \to (-t - \overline{\epsilon}, -t + \overline{\epsilon})$ such that for all $\tilde{\psi} \in U_1 \cap A$,

$$pF_A(s, \tilde{\psi}) \in T \quad \text{and} \quad |s - (-t)| < \overline{\epsilon} \quad \text{if and only if} \quad s = \overline{\sigma}(\tilde{\psi}).$$

Choose an open neighbourhood U_2 of ψ in C so small that $U_2 \subset U_1, U_2 \subset \{\tilde{\psi} \in C : \tilde{\psi}(0) > 0\}$,

$$pF_A(\overline{\sigma}(\tilde{\psi}), \tilde{\psi}) \in N_1 \quad \text{and} \quad |\overline{\sigma}(\tilde{\psi}) - (-t)| < \epsilon \quad \text{for all} \quad \tilde{\psi} \in U_2 \cap A.$$

The equation $R(\tilde{\psi}) = pF_A(\overline{\sigma}(\tilde{\psi}), \tilde{\psi})$ defines a continuous map from $U_2 \cap A$ into $N_1 \cap pA \cap T$. Next, choose an open neighbourhood $N \subset N_1$ of ρ in L so small that for all $\tilde{\rho} \in N \cap pA$,

$$F_A(\sigma(\tilde{\rho}), \tilde{\rho} + a(\tilde{\rho})) \in U_2 \quad \text{and} \quad |\sigma(\tilde{\rho}) - t| < \overline{\epsilon}.$$

The equation $Y(\tilde{\rho}) = F_A(\sigma(\tilde{\rho}), \tilde{\rho} + a(\tilde{\rho}))$ defines a continuous map from $N_1 \cap pA$ into $A \cap H$ so that $Y(N \cap pA) \subset U_2 \cap A \cap H$.

2. Claim: For $\tilde{\psi} \in U_2 \cap A \cap H$,

$$R(\tilde{\psi}) \in N_1 \cap pA \cap T \quad \text{and} \quad Y(R(\tilde{\psi})) = \tilde{\psi}.$$

Proof. Set $\tilde{\phi} = F_A(\overline{\sigma}(\tilde{\psi}), \tilde{\psi}), \tilde{\rho} = p\tilde{\phi}$. Then $\tilde{\rho} = R(\tilde{\psi}) \in N_1 \cap pA$. For $s = -\overline{\sigma}(\tilde{\psi})$, we have

$$F_A(s, \tilde{\rho} + a(\tilde{\rho})) = F_A(-\overline{\sigma}(\tilde{\psi}), F_A(\overline{\sigma}(\tilde{\psi}), \tilde{\psi})) = \tilde{\psi} \in H,$$

and

$$|-\overline{\sigma}(\tilde{\psi}) - t| = |\overline{\sigma}(\tilde{\psi}) - (-t)| < \epsilon,$$

hence $-\overline{\sigma}(\tilde{\psi}) = \sigma(\tilde{\rho})$, and therefore

$$Y(R(\tilde{\psi})) = Y(\tilde{\rho}) = F_A(\sigma(\tilde{\rho}), \tilde{\rho} + a(\tilde{\rho})) = F_A(\sigma(\tilde{\rho}), \tilde{\phi})$$

$$= F_A(-\overline{\sigma}(\tilde{\psi}), F_A(\overline{\sigma}(\tilde{\psi}), \tilde{\psi})) = \tilde{\psi}.$$

3. Claim: For $\tilde{\rho} \in N \cap pA \cap T$,

$$Y(\tilde{\rho}) \in U_2 \cap A \cap H \quad \text{and} \quad R(Y(\tilde{\rho})) = \tilde{\rho}.$$

Proof. Set $\tilde{\phi} = \tilde{\rho} + a(\tilde{\rho}), \tilde{\psi} = F_A(\sigma(\tilde{\rho}), \tilde{\phi})$. Then $\tilde{\psi} \in U_2 \cap A$. For $s = -\sigma(\tilde{\rho})$, we have

$$pF_A(s, \tilde{\psi}) = pF_A(-\sigma(\tilde{\rho}), F_A(\sigma(\tilde{\rho}), \tilde{\phi})) = p\tilde{\phi} = \tilde{\rho} \in T,$$

and

$$|-\sigma(\tilde{\rho}) - (-t)| = |\sigma(\tilde{\rho}) - t| < \overline{\epsilon},$$

hence $-\sigma(\tilde{\rho}) = \overline{\sigma}(\tilde{\psi})$. Therefore

$$R(Y(\tilde{\rho})) = R(\tilde{\psi}) = pF_A(\overline{\sigma}(\tilde{\psi}), \tilde{\psi})$$

$$= pF_A(-\sigma(\tilde{\rho}), F_A(\sigma(\tilde{\rho}), \tilde{\phi})) = p\tilde{\phi} = \tilde{\rho}.$$

4. Claim: $Y(N \cap pA \cap T) = (R^{-1}(N \cap pA)) \cap H$. Proof. Consider $\tilde{\psi} = Y(\tilde{\rho})$ where $\tilde{\rho} \in N \cap pA \cap T$. By the preceding claim,

$$\tilde{\psi} \in U_2 \cap A \cap H, \quad \text{and} \quad R(\tilde{\psi}) = R(Y(\tilde{\rho})) = \tilde{\rho} \in N \cap pA.$$

In particular, $\tilde{\psi} \in R^{-1}(N \cap pA)$, and we get $\tilde{\psi} \in (R^{-1}(N \cap pA)) \cap H$. If $\tilde{\psi} \in (R^{-1}(N \cap pA)) \cap H$, then $\tilde{\psi} \in U_2 \cap A \cap H$, and part 2 of the proof yields $R(\tilde{\psi}) \in N_1 \cap pA \cap T$ and $Y(R(\tilde{\psi})) = \tilde{\psi}$. Since $R(\tilde{\psi}) \in N \cap pA$, we get $\tilde{\psi} \in Y(N \cap pA \cap T)$.

5. The set $R^{-1}(N \cap pA)$ is an open neighbourhood of x_t in the space A, and is contained in $U_2 \cap A$. It follows that there is an open set U of C so that $x_t \in U \subset U_2$ and $(R^{-1}(N \cap pA)) \cap H = U \cap A \cap H$. Furthermore, we see that Y maps the open neighbourhood $N \cap pA \cap T$ of ρ in the space $pA \cap T$ homeomorphically onto the open neighbourhood $U \cap A \cap H$ of x_t in the space $A \cap H$.

6. Proof of $U \cap A \cap H = U \cap A(P)$. By Proposition 4.4, $A(P) = \overline{K} \cap A$. Because of $\overline{K} = K \cup \{0\}$ and $0 \notin U$ it remains to show that $U \cap A \cap H = U \cap A \cap K$. For $\psi \in U \cap A \cap H$, we have $\psi(-1) = 0 < \psi(0)$. It follows that $x(\psi)$ is slowly oscillating, and $\psi = x(\psi)_0 \in K$ (see Chapter 3). — The other inclusion is trivial. $\square$

COROLLARY 8.2. *Let $\phi \in A \setminus \{0\}$ be given. Let $\rho = p\phi, x = x(\phi), t \in \mathbb{R}$. Suppose the line $T = \rho + \mathbb{R}\chi, 0 \neq \chi \in L$, is transversal to the projected phase curve $s \mapsto px_s$ at $s = 0$, and the line $\tilde{T} = \tilde{\rho} + \mathbb{R}\tilde{\chi}, 0 \neq \tilde{\chi} \in L$ and $\tilde{\rho} = px_t$, is transversal to $s \mapsto px_s$ at $s = t$. Then there exist open neighbourhoods N of ρ in L and $\tilde{N}$ of $\tilde{\rho}$ in L, $\epsilon > 0$, and a continuous map $\sigma : N \cap pA \to (t - \epsilon, t + \epsilon)$ with $\sigma(\rho) = t$ such that the map*

$$N \cap pA \cap T \ni \overline{\rho} \mapsto pF_A(\sigma(\overline{\rho}), \overline{\rho} + a(\overline{\rho})) \in pA$$

defines a homeomorphism onto the set $\tilde{N} \cap pA \cap \tilde{T}$.

SKETCH OF THE PROOF. Proposition 8.1(ii) yields an open neighbourhood U of ϕ in C, $\epsilon > 0$, and a continuous map $\sigma : U \cap A \to (t - \epsilon, t + \epsilon)$ such that for all $\overline{\phi} \in U \cap A$,

$$pF_A(s, \overline{\phi}) \in \tilde{T} \quad \text{and} \quad |s - t| < \epsilon \quad \text{if and only if} \quad s = \sigma(\overline{\phi}).$$

The set $p(U \cap A)$ is an open neighbourhood of ρ in the space pA, and there exists an open neighbourhood N_1 of ρ in the space L with $p(U \cap A) = N_1 \cap pA$. For all $\overline{\rho} \in N_1 \cap pA$ we have

$$pF_A(s, \overline{\rho} + a(\overline{\rho})) \in \tilde{T} \quad \text{and} \quad |s - t| < \epsilon \quad \text{if and only if} \quad s = \sigma(\overline{\rho} + a(\overline{\rho})).$$

Analogously one finds an open neighbourhood $\tilde{N}_1$ of $\tilde{\rho} = px_t$ in L, $\tilde{\epsilon} > 0$, and a continuous map $\tilde{\sigma} : \tilde{N}_1 \cap pA \to (-t - \tilde{\epsilon}, -t + \tilde{\epsilon})$ such that for all $\rho^* \in \tilde{N}_1 \cap pA$ we have

$$pF_A(s, \rho^* + a(\rho^*)) \in T \quad \text{and} \quad |s - (-t)| < \tilde{\epsilon} \quad \text{if and only if} \quad s = \tilde{\sigma}(\rho^*).$$

From here, one proceeds as in the proof of Corollary 8.1. □

CHAPTER 9

Angles Along Projected Phase Curves

Let a slowly oscillating solution $x : [t_0 - 1, \infty) \to \mathbb{R}$ of equation (1.1) be given. Suppose the zeros in the interval (t_0, ∞) form a sequence $(z_j)_0^J$, $J \in \mathbb{N}$ or $J = \infty$, with property (so) and with $x'(z_0) > 0$. The phase curve $X : [t_0, \infty) \ni t \mapsto x_t \in C$ has values in S. Lemma 5.1 gives $|p \circ X| \subset L \setminus \{0\}$. We are interested in lower estimates for angles along the smooth curve $p \circ X$. Recall from Chapter 5 the basis $\{\beta_1, \beta_2\}$ of the space L and the isomorphism I onto the $\mathbb{R}$–vectorspace $\mathbb{C}$. Choose an argument $\xi_0 \in \mathbb{R}$ for the point $I \circ p \circ X(z_0 + 1) \in \mathbb{C}$, i.e.,

$$I \circ p \circ X(z_0 + 1) = |I \circ p \circ X(z_0 + 1)| e^{i\xi_0},$$

and set

$$\xi(t) = \Re \frac{1}{i} \int_{I \circ p \circ X|[z_0+1,t]} \frac{1}{z} dz + \xi_0 \quad \text{for} \quad t \geq z_0 + 1.$$

The set

$$K_0 = \{\phi \in C : 0 < \phi(t) \quad \text{for} \quad -1 < t < 0\}$$

is a convex cone contained in S.

PROPOSITION 9.1. *Suppose $t_0 \leq s, s + 1 \leq t$, and that x has no zero in the interval $(s, t]$. Then $|\xi(t) - \xi(s)| < \pi$.*

PROOF. In case $x(u) > 0$ for $s < u < t$ we have $x_u \in K_0$ for $s + 1 \leq u \leq t$. Hence $I \circ p \circ X(u) \in I \circ p(K_0) \subset \mathbb{C} \setminus \{0\}$ for such u. By Lemma 2.1, $I \circ p \circ X(s) = |I \circ p \circ X(s)| e^{i\xi(s)}$. The set $I \circ p(K_0)$ is a convex cone. Apply Corollary 2.1. In case $x(u) < 0$ for $s < u < t$, use the cone $-K_0$. $\square$

Lower estimates for the angle ξ on intervals in which x has zeros are more difficult to obtain. We shall employ a comparison curve for this, namely the simple closed piecewise smooth curve $c : [0, b] \to L$ which is defined in [20, Section 4]. The definition depends on the location of the leading eigenvalues of the generator of the semigroup of the linear equation (1.2). For example, in case III of Chapter 5, i.e., if the leading eigenvalues form a complex conjugate pair $u_0 + iv_0, u_0 - iv_0$, with $0 < v_0 < \pi$, we have

$$c(t) = \sin(v_0(t + \cdot + 1)) \in L \quad \text{for} \quad 0 \leq t \leq \frac{2\pi}{v_0}.$$

The function $t \mapsto e^{u_0(t+1)}\sin(v_0(t+1))$ is a slowly oscillating solution of equation (1.2), with consecutive zeros at $-1, \frac{\pi}{v_0} - 1, \frac{2\pi}{v_0} - 1$. The closed curve c starts in the cone K, intersects $-K$ at $t = \frac{\pi}{v_0}$, and returns to K. In the cases I and II of Chapter 5 the curve c is a suitable parameterization of the boundary of the quadrangle

$$\{r\beta_1 + s\beta_2 : |r| \le 1, |s| \le 1\} \subset L.$$

There exists an argument $\zeta_0 \in (\xi_0 - \pi, \xi_0 + \pi]$ of the point $I \circ c(0) \in \mathbb{C}$, i.e.,

$$I \circ c(0) = |I \circ c(0)|e^{i\zeta_0},$$

and the angle

$$\zeta(t) = \Re\frac{1}{i}\int_{I \circ c|[0,t]} \frac{1}{z}dz + \zeta_0, \quad 0 \le t \le b,.$$

along c is strictly increasing, with

$$\zeta(0) = \zeta_0 \quad \text{and} \quad \zeta(b) = 2\pi + \zeta(0).$$

This follows easily from the definition of the curve c. For details in the cases I and II, see [20, Section 4].

In order to compare ξ to ζ we need homotopies in L which avoid the origin. The observation which leads to suitable homotopies is that the curves $c = p \circ c$ and X travel through finite families of cones in S which are convex (Recall that the positively invariant set S itself is not convex). More precisely, the situation is as follows. For every integer $k \ge 3$, consider the $k+1$ convex sets

$$K_{k0} = K_0 = \{\phi \in C : 0 < \phi(t) \quad \text{for} \quad -1 < t < 0\},$$

$$K_{k1} = \{\phi \in C : 0 < \phi(t) \quad \text{for} \quad -1 \le t < \frac{1}{k},$$

$$\phi'(t) \quad \text{exists and} \quad \phi'(t) < 0 \quad \text{for} \quad -\frac{1}{k} \le t \le 0\},$$

for $\kappa = 2, \ldots, k-1$,

$$K_{k\kappa} = \{\phi \in C : 0 < \phi(t) \quad \text{for} \quad -1 \le t < -\frac{\kappa}{k},$$

$$\phi'(t) \quad \text{exists and} \quad \phi'(t) < 0 \quad \text{for} \quad -\frac{\kappa}{k} \le t \le -\frac{\kappa-1}{k},$$

$$\phi(t) < 0 \quad \text{for} \quad -\frac{\kappa-1}{k} < t \le 0\},$$

$$K_{kk} = \{\phi \in C : \phi'(t) \quad \text{exists and} \quad \phi'(t) < 0 \quad \text{for} \quad -1 \le t \le -1 + \frac{1}{k},$$

$$\phi(t) < 0 \quad \text{for} \quad -1 + \frac{1}{k} < t \le 0\}.$$

All sets $K_{k\kappa}$ are contained in S. The definition of the curve c implies that there exists an integer $k(c) \ge 3$ such that for every integer $k \ge k(c)$ there is a subdivision

(9.1) $0 = t_0 < t_1 < \cdots < t_{2k+2} = b$

so that

$$(9.2) \qquad c(t) \in K_{k\kappa} \quad \text{for} \quad t_\kappa \leq t \leq t_{\kappa+1} \quad \text{and} \quad \kappa = 0, \ldots, k,$$

$$c(t) \in -K_{k\kappa} \quad \text{for} \quad t_{k+1+\kappa} \leq t \leq t_{k+1+\kappa+1} \quad \text{and} \quad \kappa = 0, \ldots, k.$$

For details in the cases I and II, see [**20**, Section 4].

Now consider the slowly oscillating solution $x : [t_0 - 1, \infty) \to \mathbb{R}$. There exists an integer $k(x, z_0) \geq 3$ such that for every integer $k \geq k(x, z_0)$ there is a subdivision

$$(9.3) \qquad z_0 + 1 = s_0 < s_1 < \cdots < s_{k+1} = z_1 + 1$$

so that

$$(9.4) \qquad x_s \in K_{k\kappa} \quad \text{for} \quad s_\kappa \leq s \leq s_{\kappa+1} \quad \text{and} \quad \kappa = 0, \ldots, k.$$

If $J \geq 2$ then there is a next zero $z_2 > z_1$, and there is an integer $k(x, z_1) \geq k(x, z_0)$ such that for every integer $k \geq k(x, z_1)$ there exists a subdivision

$$(9.5) \qquad z_1 + 1 = s_{k+1} < \cdots < s_{2k+2} = z_2 + 1$$

so that

$$(9.6) \quad x_s \in -K_{k\kappa} \quad \text{for} \quad s_{k+1+\kappa} \leq s \leq s_{k+1+\kappa+1} \quad \text{and} \quad \kappa = 0, \ldots, k.$$

PROPOSITION 9.2. *There exists a strictly increasing function* $\beta : [z_0 + 1, z_1 + 1] \to \mathbb{R}$ *with* $\beta(z_0 + 1) = \zeta_0$ *and*

$$|\xi(t) - \beta(t)| < \pi \quad \text{for} \quad z_0 + 1 \leq t \leq z_1 + 1.$$

In case $J \geq 2$ *there is a strictly increasing function* $\beta : [z_0 + 1, z_2 + 1] \to \mathbb{R}$ *with* $\beta(z_0 + 1) = \zeta_0, \beta(z_2 + 1) = 2\pi + \beta(z_0 + 1)$ *and*

$$|\xi(t) - \beta(t)| < \pi \quad \text{for} \quad z_0 + 1 \leq t \leq z_2 + 1.$$

PROOF. 1. Suppose $J \geq 2$. We derive the last assertion of Proposition 9.2. Consider $\zeta : [0, b] \to \mathbb{R}$ as before, with $\xi_0 - \pi < \zeta_0 \leq \xi_0 + \pi$. Fix an integer $k \geq k(c)$ and let subdivisions as in (9.3) and (9.5) be given so that (9.4) and (9.6) hold. Recall the subdivision (9.1). We reparameterize c. Let a_κ denote the affine linear map given by

$$a_\kappa(s_{\kappa-1}) = t_{\kappa-1}, \quad a_\kappa(s_\kappa) = t_\kappa, \quad \text{for} \quad \kappa = 1, \ldots, 2k+2.$$

For $s_{\kappa-1} \leq s \leq s_\kappa$ and $\kappa = 1, \ldots, 2k+2$, set

$$c_r(s) = c(a_\kappa(s)).$$

This yields a piecewise smooth curve in $L \setminus \{0\}$, and the angle

$$\beta(t) = \Re \frac{1}{i} \int_{I \circ c_r | [z_0+1, t]} \frac{1}{z} dz + \zeta_0, \quad z_0 + 1 \leq t \leq z_2 + 1,$$

is strictly increasing. We have $\beta(z_0 + 1) = \zeta_0$.

2. The curve c_r is simple and closed, and it satisfies

$$\beta(z_2 + 1) = 2\pi + \beta(z_0 + 1).$$

We have

$$c_r(s) \in K_{k\kappa} \quad \text{and} \quad X(s) \in K_{k\kappa} \quad \text{for} \quad s_\kappa \leq s \leq s_{\kappa+1} \quad \text{and} \quad \kappa = 0, \ldots, k$$

and

$$c_r(s) \in -K_{k\kappa} \quad \text{and} \quad X(s) \in -K_{k\kappa} \quad \text{for} \quad s_{k+1+\kappa} \leq s \leq s_{k+1+\kappa+1} \quad \text{and}$$

$$\kappa = 0, \ldots, k.$$

3. Claim: For $z_0 + 1 \leq s \leq z_2 + 1$ and $0 \leq t \leq 1$,

$$t c_r(s) + (1 - t) X(s) \in S.$$

Proof. Given s and t, there exists $\kappa \in \{0, \ldots, k\}$ so that either $c_r(s) \in K_{k\kappa} \ni X(s)$, or $c_r(s) \in -K_{k\kappa} \ni X(s)$. The cones $K_{k\kappa}, -K_{k\kappa}$ are convex and contained in the set S.

4. Claim: $\zeta_0 < \xi_0 + \pi$. Proof. Recall $c_r(z_0 + 1) = c(0) \in K_{k0} \ni X(z_0 + 1)$. Consequently, both points $p \circ c_r(z_0 + 1) = c_r(z_0 + 1)$ and $p \circ X(z_0 + 1)$ belong to the convex cone $pK_{k0} \subset pS \subset L \setminus \{0\}$. Therefore, the trace of the map

$$w : [0, 1] \ni t \mapsto tI \circ c_r(z_0 + 1) + (1 - t)I \circ p \circ X(z_0 + 1) \in \mathbb{C}$$

is a line segment in a convex cone in $\mathbb{C} \setminus \{0\}$. We have $w(0) = |w(0)|e^{i\xi_0}$. Corollary 2.2 implies that the angle γ along w with $\gamma(0) = \xi_0$ satisfies

$$\pi > |\gamma(1) - \gamma(0)| = |\gamma(1) - \xi_0|.$$

Lemma 2.1 gives

$$|I \circ c(0)|e^{i\zeta_0} = I \circ c(0) = I \circ c_r(z_0 + 1) = w(1) = |w(1)|e^{i\gamma(1)}.$$

Both ζ_0 and $\gamma(1)$ are contained in the interval $(\xi_0 - \pi, \xi_0 + \pi]$. It follows that $\zeta_0 = \gamma(1)$. Finally,

$$|\zeta_0 - \xi_0| = |\gamma(1) - \gamma(0)| < \pi.$$

5. Construction of a homotopy $h : [0, 1] \times [-2, 4] \to L$ of closed piecewise smooth curves. Let $\tilde{a}$ denote the affine linear map which satisfies $\tilde{a}(1) = z_0 + 1, \tilde{a}(2) = z_2 + 1$. For $\theta \in [0, 1]$ we define

$$
\begin{aligned}
h(\theta, t) &= \frac{1}{2}(1 - t)p \circ X \circ \tilde{a}(1) + \frac{1}{2}(t + 1)c_r \circ \tilde{a}(1) \quad \text{for} \quad -1 \leq t \leq 1, \\
h(\theta, t) &= p \circ X \circ \tilde{a}(1 + \theta(-1 - t)) \quad \text{for} \quad -2 \leq t \leq -1, \\
h(\theta, t) &= c_r \circ \tilde{a}(1 + \theta(t - 1)) \quad \text{for} \quad 1 \leq t \leq 2, \\
h(\theta, t) &= \frac{1}{2}(t - 2)h(\theta, 2) + \frac{1}{2}(4 - t)h(\theta, -2) \quad \text{for} \quad 2 \leq t \leq 4.
\end{aligned}
$$

Note

$$h(\theta, -2) = p \circ X \circ \tilde{a}(1 + \theta), \quad h(\theta, 2) = c_r \circ \tilde{a}(1 + \theta).$$

I.e., given $\theta \in [0,1]$, we follow the trace $|p \circ X|$ from $p \circ X(\tilde{a}(1+\theta))$ back to $p \circ X(\tilde{a}(1)) = p \circ X(z_0+1)$, then go along a straight line from $p \circ X(z_0+1)$ to $c_r(z_0+1)$, proceed along the trace $|c_r|$ to the point $c_r(\tilde{a}(1+\theta))$, and return back to the point $p \circ X(\tilde{a}(1+\theta))$ along a straight line.

Recall from part 3 that the traces $|X|, |c_r|$, and the line segments which connect $X(z_0+1)$ to $c_r(z_0+1)$ and $c_r(\tilde{a}(1+\theta))$ to $X(\tilde{a}(1+\theta))$ all belong to the set S. Using Lemma 5.1 and $p \circ c_r = c_r$ we infer

$$h([0,1] \times [-2,4]) \subset L \setminus \{0\}.$$

Therefore the winding numbers of the closed piecewise smooth curves $h(\theta, \cdot)$, $0 \le \theta \le 1$, with respect to the origin $0 \in L$ are defined. The trace $|h(0,\cdot)|$ is a line segment in $L \setminus \{0\}$. It follows that

(9.7) $\mathrm{wind}(0, h(0,\cdot)) = 0.$

We show that h is continuous. Continuity on the open subset $[0,1] \times ([-2,-1) \cup (-1,1) \cup (1,2))$ of its domain $[0,1] \times [-2,4]$ is obvious from the formulae defining h. On the set $[0,1] \times [2,4]$ we have

$$h(\theta, t) = \frac{1}{2}(t-2)c_r(\tilde{a}(1+\theta)) + \frac{1}{2}(4-t)p \circ X(\tilde{a}(1+\theta)).$$

This yields continuity on the open subset $[0,1] \times (2,4]$. Next, consider a point $(\theta, 2), 0 \le \theta \le 1$, and a sequence of points $(\theta_n.t_n) \in [0,1] \times [-2,4]$ converging to it. For every subsequence of indices n_ν with $1 \le t_{n_\nu} \le 2$ for all ν,

$$h(\theta_{n_\nu}, t_{n_\nu}) = c_r \circ \tilde{a}(1 + \theta_{n_\nu}(t_{n_\nu} - 1)) \to c_r \circ \tilde{a}(1+\theta) = h(\theta, 2)$$

as $\nu \to \infty$. For every subsequence of indices n_ν with $t_{n_\nu} \ge 2$ for all ν,

$$h(\theta_{n_\nu}, t_{n_\nu}) = \frac{1}{2}(t_{n_\nu} - 2)c_r \circ \tilde{a}(1 + \theta_{n_\nu}) + \frac{1}{2}(4 - t_{n_\nu})p \circ X \circ \tilde{a}(1 + \theta_{n_\nu})$$

$$\to \frac{1}{2}(t-2)c_r \circ \tilde{a}(1+\theta) + \frac{1}{2}(4-t)p \circ X \circ \tilde{a}(1+\theta) = h(\theta, t)$$

as $\nu \to \infty$. It follows that $h(\theta_n, t_n) \to h(\theta, t)$ as $n \to \infty$, and h is continuous at the point (θ, t). Now consider a point $(\theta, -1), 0 \le \theta \le 1$, a sequence of points $(\theta_n, t_n) \in [0,1] \times (-1,2)$ which converges to $(\theta, -1)$ as $n \to \infty$, and subsequences of points $(\theta_{n_\nu}, t_{n_\nu})$ so that $t_{n_\nu} \le -1$ for all ν; $(\theta_{n_\iota}, t_{n_\iota})$ so that $t_{n_\iota} \ge -1$ for all ι. Then

$$h(\theta_{n_\nu}, t_{n_\nu}) = p \circ X \circ \tilde{a}(1 + \theta_{n_\nu}(-1 - t_{n_\nu}))$$

$$\to p \circ X \circ \tilde{a}(1 + \theta(-1 - (-1))) = p \circ X(\tilde{a}(1)) = h(\theta, -1) \quad \text{as} \quad \nu \to \infty,$$

and

$$h(\theta_{n_\iota}, t_{n_\iota}) = \frac{1}{2}(1 - t_{n_\iota})p \circ X(\tilde{a}(1)) + \frac{1}{2}(1 + t_{n_\iota})c_r(\tilde{a}(1))$$

$$\to p \circ X(\tilde{a}(1)) = h(\theta, -1) \quad \text{as} \quad \iota \to \infty;$$

we obtain $h(\theta_n, t_n) \to h(\theta, -1)$ as $n \to \infty$, and h is continuous at $(\theta, -1)$. The proof for points $(\theta, 1), 0 \le \theta \le 1$, is analogous.

6. The homotopy invariance of the winding number and (9.7) imply that for every $\theta \in [0,1]$,

$$0 = \text{wind}(0, h(\theta, \cdot)) = \Re \frac{1}{2\pi i} \int_{Ioh(\theta,\cdot)} \frac{1}{z} dz.$$

Therefore we have

(9.8) $$0 = \Re \frac{1}{i} \int_{Ioh(\theta,\cdot)|[-2,-1]} \frac{1}{z} dz + \Re \frac{1}{i} \int_{Ioh(\theta,\cdot)|[-1,1]} \frac{1}{z} dz$$

$$+ \Re \frac{1}{i} \int_{Ioh(\theta,\cdot)|[1,2]} \frac{1}{z} dz + \Re \frac{1}{i} \int_{Ioh(\theta,\cdot)|[2,4]} \frac{1}{z} dz.$$

The third term in the last sum equals

$$\Re \frac{1}{i} \int_{Ioc_r|[z_0+1,z_0+1+\theta(z_2-z_0)]} \frac{1}{z} dz = \beta(z_0 + 1 + \theta(z_2 - z_0)) - \zeta_0.$$

The first term in the last sum equals

$$\Re(-\frac{1}{i} \int_{IopoX|[z_0+1,z_0+1+\theta(z_2-z_0)]} \frac{1}{z} dz)$$

$$= -(\xi(z_0 + 1 + \theta(z_2 - z_0)) - \xi_0) = \xi_0 - \xi(z_0 + 1 + \theta(z_2 - z_0)).$$

Corollary 2.2 implies that the second and fourth term in the sum in (9.8) have absolute values strictly less than π.
Claim:

$$\Re \frac{1}{i} \int_{Ioh(\theta,\cdot)|[-1,1]} \frac{1}{z} dz = \zeta_0 - \xi_0.$$

Proof. Recall

$$I \circ h(\theta, -1) = I \circ p \circ X(z_0 + 1) = |I \circ p \circ X(z_0 + 1)|e^{i\xi_0}.$$

Lemma 2.1 and Corollary 2.2, applied to

$$\delta(t) = \Re \frac{1}{i} \int_{Ioh(\theta,\cdot)|[-1,t]} \frac{1}{z} dz + \xi_0,$$

yield

$$I \circ h(\theta, 1) = |I \circ h(\theta, 1)|e^{i\delta(1)} \quad \text{and} \quad |\delta(1) - \xi_0| < \pi.$$

Recall

$$I \circ h(\theta, 1) = I \circ c_r(z_0 + 1) = |I \circ c_r(z_0 + 1)|e^{i\zeta_0},$$

and $|\zeta_0 - \xi_0| < \pi$ (see part 4). Now it becomes obvious that $\delta(1) = \zeta_0$.
7. For $z_0 + 1 \le t \le z_2 + 1$ we have $t = z_0 + 1 + \theta(z_2 - z_0)$ with $\theta \in [0,1]$. Using the results of part 6 we obtain

$$|\xi(t) - \beta(t)| = |\beta(z_0 + 1 + \theta(z_2 - z_0)) - \xi(z_0 + 1 + \theta(z_2 - z_0))|$$

$$= |\Re \frac{1}{i} \int_{Ioc_r|[z_0+1,z_0+1+\theta(z_2-z_0)]} \frac{1}{z} dz + \zeta_0$$

$$+\Re(-\frac{1}{i}\int_{I\circ po X\,|[z_0+1,z_0+1+\theta(z_2-z_0)]}\frac{1}{z}dz)-\xi_0|$$

$$=|\Re\frac{1}{i}\int_{I\circ h(\theta,\cdot)|[1,2]}\frac{1}{z}dz+\Re\frac{1}{i}\int_{I\circ h(\theta,\cdot)|[-2,-1]}\frac{1}{z}dz+\Re\frac{1}{i}\int_{I\circ h(\theta,\cdot)|[-1,1]}\frac{1}{z}dz|$$

$$=|-\Re\frac{1}{i}\int_{I\circ h(\theta,\cdot)|[2,4]}\frac{1}{z}dz| \quad \text{(see (9.8))}$$

$$< \pi.$$

8. On the first assertion of Proposition 9.2. In this case, fix an integer $k \geq k(c)$. There are subdivisions (9.1) and (9.3) so that (9.2) and (9.5) hold. For $\kappa = 1,\ldots,k+1$, define affine linear maps a_κ by

$$a_\kappa(s_{\kappa-1}) = t_{\kappa-1}, \quad a_\kappa(s_\kappa) = t_\kappa.$$

Define a curve c_ρ by

$$c_\rho(s) = c(a_\kappa(s)) \quad \text{for} \quad s_{\kappa-1} \leq s \leq s_\kappa \quad \text{and} \quad \kappa = 1,\ldots,k+1.$$

(This is not a closed curve, in contrast to the curve c_r used before.) Then we have

$$c_\rho(s) \in K_{k\kappa} \ni X(s) \quad \text{for} \quad s_{\kappa-1} \leq s \leq s_\kappa \quad \text{and} \quad \kappa = 1,\ldots,k+1,$$

and the angle β along c_ρ with $\beta(z_0 + 1) = \zeta_0$ is strictly increasing. The rest of the proof is analogous to the case $J \geq 2$. $\square$

COROLLARY 9.1. *In case $J = \infty$ there exists a strictly increasing function $\beta : [z_0+1,\infty) \to \mathbb{R}$ with $\beta(z_0+1) = \zeta_0$,*

$$|\xi(t) - \beta(t)| < \pi \quad \text{for} \quad z_0+1 \leq t,$$

and

$$\beta(z_{2j}+1) = 2\pi j + \zeta_0 \quad \text{for all integers} \quad j \geq 0.$$

In case $J \in \mathbb{N}$ there exists a strictly increasing function $\beta : [z_0+1, z_J+1] \to \mathbb{R}$ with $\beta(z_0+1) = \zeta_0$,

$$|\xi(t) - \beta(t)| < \pi \quad \text{for} \quad z_0+1 \leq t \leq z_J+1,$$

and

$$\beta(z_{2j}+1) = 2\pi j + \zeta_0 \quad \text{for all} \quad j \in \mathbb{N}_0 \quad \text{with} \quad j \leq \frac{J}{2}.$$

PROOF. 1. The case $J \in \mathbb{N}$. For $J = 1$ and $J = 2$, see Proposition 9.2. So assume $3 \leq J$. Let J^* denote the largest integer in $\mathbb{N}$ with $J^* \leq \frac{J}{2}$.
1.1. We proceed by induction and show that for every $j \in \mathbb{N}$ with $j \leq J^*$ there exists a strictly increasing function $\beta_{(j)} : [z_0+1, z_{2j}+1] \to \mathbb{R}$ with $\beta_{(j)}(z_0+1) = \zeta_0$,

$$|\xi(t) - \beta_{(j)}(t)| < \pi \quad \text{for} \quad z_0+1 \leq t \leq z_{2j}+1,$$
$$\beta_{(j)}(z_{2\iota}+1) = 2\pi\iota + \zeta_0 \quad \text{for} \quad \iota = 1,\ldots,j.$$

For $j = 1$, see Proposition 9.2. Suppose the assertion holds true for an integer $j \in \{1, \ldots, J^* - 1\}$. Note $x'(z_{2j}) > 0$. Lemma 2.1 yields

$$I \circ p \circ X(z_{2j} + 1) = |I \circ p \circ X(z_{2j} + 1)|e^{i\xi(z_{2j}+1)}.$$

For $z_{2j} + 1 \leq t \leq z_{2j+2} + 1$ we have

$$\xi(t) = \Re \frac{1}{i} \int_{I \circ p \circ X|[z_{2j}+1,t]} \frac{1}{z} dz + \xi(z_{2j} + 1).$$

Apply the second statement of Proposition 9.2 to the zero z_{2j} (instead of z_0), and to the argument $\xi^* = \xi(z_{2j} + 1)$ (instead of ξ_0). For $\zeta^* \in (\xi^* - \pi, \xi^* + \pi]$ with

$$(9.9) \qquad\qquad I \circ c(0) = |I \circ c(0)|e^{i\zeta^*}$$

we obtain a strictly increasing function

$$\beta^* : [z_{2j} + 1, z_{2j+2} + 1] \to \mathbb{R}$$

with

$$(9.10) \qquad\qquad \beta^*(z_{2j} + 1) = \zeta^*$$

and

$$\beta^*(z_{2j+2} + 1) = 2\pi + \beta^*(z_{2j} + 1)$$

such that

$$(9.11) \qquad |\xi(t) - \beta^*(t)| < \pi \quad \text{for} \quad z_{2j} + 1 \leq t \leq z_{2j+2} + 1.$$

From the induction hypothesis,

$$(9.12) \qquad\qquad \pi > |\xi(z_{2j} + 1) - \beta_{(j)}(z_{2j} + 1)|,$$

$$(9.13) \qquad \beta_{(j)}(z_{2j} + 1) = 2\pi j + \beta_{(j)}(z_0 + 1) = 2\pi j + \zeta_0.$$

Recall $I \circ c(0) = |I \circ c(0)|e^{i\zeta_0}$. Using (9.13) we get

$$I \circ c(0) = |I \circ c(0)|e^{i\beta_{(j)}(z_{2j}+1)}.$$

From (9.9) and (9.10),

$$I \circ c(0) = |I \circ c(0)|e^{i\beta^*(z_{2j}+1)}.$$

Using (9.11) and (9.12) we infer from the last equations that

$$\beta_{(j)}(z_{2j} + 1) = \beta^*(z_{2j} + 1).$$

Set

$$\beta_{(j+1)}(t) = \beta_{(j)}(t) \quad \text{for} \quad z_0 + 1 \leq t \leq z_{2j} + 1,$$
$$\beta_{(j+1)}(t) = \beta^*(t) \quad \text{for} \quad z_{2j} + 1 \leq t \leq z_{2j+2} + 1.$$

This yields a strictly increasing function $\beta_{(j+1)}$ with the desired properties.

1.2. In case $J = 2J^*$, set $\beta = \beta_{(J^*)}$. In case $J = 2J^* + 1$, we apply the first assertion of Proposition 9.2 to the zero z_{2J^*} and to the argument $\xi(z_{2J^*} + 1)$. This yields an angle function $\overline{\beta}$ on the interval $[z_{2J^*} + 1, z_{2J} + 1]$, and we construct β from $\beta_{(J^*)}$ and $\overline{\beta}$ as in part 1.1.

2. The case $J = \infty$. As in part 1.1 we obtain functions $\beta_{(j)}$ so that $\beta_{(j+1)}(t) = \beta_{(j)}(t)$ for all $t \in [z_0 + 1, z_{2j} + 1]$ and all $j \in \mathbb{N}$. Set $\beta(t) = \beta_{(j)}(t)$ for $z_{2j-2} + 1 \leq t < z_{2j} + 1$ and $j \in \mathbb{N}$. $\square$

The proofs of the preceding comparison results did not make use of the Chapters 4,6,7, and 8. We end this section with corollaries concerning periodic solutions. The first of these sharpens [**20**, Theorem 5.1]. The second one corresponds to [**20**, Corollary 5.1]. The proofs rely on the fact that the projection p is injective on unions of orbits in C of slowly oscillating periodic solutions, i.e., on Theorem 7.1(i).

COROLLARY 9.2. *Suppose x is the restriction of a periodic solution. Then the minimal period τ of x equals $z_2 - z_0$, and*

$$|\Re\frac{1}{i}\int_{I\circ p\circ X|[s,t]}\frac{1}{z}dz| < 4\pi \quad for \quad z_0 + 1 \leq s < t < z_2 + 1.$$

For the simple closed curve $\eta : [0,\tau] \ni t \mapsto x_t \in C$, we have $0 \in \mathrm{int}(p \circ \eta)$.

PROOF. Let $\tau > 0$ denote the minimal period of x. We have $J = \infty$, and property (so) yields $\tau = z_{2n} - z_0$ for some $n \in \mathbb{N}$. Theorem 7.1(i) implies that the restriction $I \circ p \circ X|[z_0 + 1, z_{2n} + 1]$ is a simple closed smooth curve in $\mathbb{C} \setminus \{0\}$. Therefore

$$\mathrm{wind}(0, I \circ p \circ X|[z_0 + 1, z_{2n} + 1]) \in \{-1, 0, 1\},$$

and

$$
\begin{aligned}
2\pi \quad &\geq \quad \Re\frac{1}{i}\int_{I\circ p\circ X|[z_0+1,z_{2n}+1]}\frac{1}{z}dz = \xi(z_{2n} + 1) - \xi(z_0 + 1) \\
&> \quad \beta(z_{2n} + 1) - \beta(z_0 + 1) - 2\pi \quad \text{(Corollary 9.1)} \\
&= \quad (2n - 2)\pi \quad \text{(Corollary 9.1)}.
\end{aligned}
$$

This gives $n = 1$. Next, if $z_0 + 1 \leq s < t < z_2 + 1 = z_0 + 1 + \tau$, then

$$
\begin{aligned}
\Re\frac{1}{i}\int_{I\circ p\circ X|[s,t]}\frac{1}{z}dz \quad &= \quad \xi(t) - \xi(s) \\
&< \quad \beta(t) + \pi - \beta(s) + \pi \quad \text{(Corollary 9.1)} \\
&\leq \quad \beta(z_2 + 1) - \beta(z_0 + 1) + 2\pi \quad \text{(monotonicity of } \beta) \\
&= \quad 4\pi \quad \text{(Corollary 9.1)}.
\end{aligned}
$$

Analogously,

$$\begin{aligned}
\xi(t) - \xi(s) &> -\beta(t) - \pi + \beta(s) - \pi \\
&\geq -\beta(z_2 + 1) - \pi + \beta(z_0 + 1) - \pi \\
&= -4\pi.
\end{aligned}$$

The last assertion follows from

$$2\pi \cdot \mathrm{wind}(0, p \circ X \,|[z_0 + 1, z_2 + 1]) = \Re \frac{1}{i} \int_{I \circ p \circ X \,|[z_0+1, z_2+1]} \frac{1}{z} dz$$

$$= \xi(z_2 + 1) - \xi(z_0 + 1) > \beta(z_2 + 1) - \beta(z_0 + 1) - 2\pi = 0.$$

$\square$

COROLLARY 9.3. *Suppose x is the restriction of a periodic solution with minimal period $\tau > 0$, and $\tilde{x} : \mathbb{R} \to \mathbb{R}$ is a slowly oscillating periodic solution of equation (1.1) with minimal period $\tilde{\tau} > 0$. If the simple closed curves $\eta : [0, \tau] \ni t \mapsto x_t \in C$ and $\tilde{\eta} : [0, \tilde{\tau}] \ni t \mapsto \tilde{x}_t \in C$ satisfy $|\eta| \neq |\tilde{\eta}|$, then we have either $|p \circ \eta| \subset \mathrm{int}(p \circ \tilde{\eta})$, or $|p \circ \tilde{\eta}| \subset \mathrm{int}(p \circ \eta)$.*

PROOF. Suppose $|\eta| \neq |\tilde{\eta}|$. Then $|\eta| \cap |\tilde{\eta}| = \emptyset$. Both orbits belong to A. Theorem 7.1(i) implies

(9.14) $|p \circ \eta| \cap |p \circ \tilde{\eta}| = \emptyset.$

Corollary 9.1 gives $0 \in \mathrm{int}(p \circ \eta) \cap \mathrm{int}(p \circ \tilde{\eta})$. There is a line segment from 0 to a point $\chi \in |p \circ \eta|$ with minimal norm. It follows that $[0, 1)\chi \subset \mathrm{int}(p \circ \eta)$. In case $[0, 1)\chi \cap |p \circ \tilde{\eta}| = \emptyset$ we conclude that $\chi \in \mathrm{int}(p \circ \tilde{\eta})$. Using (9.14) we get $|p \circ \eta| \subset \mathrm{int}(p \circ \tilde{\eta})$. In the remaining case,

$$\emptyset \neq [0, 1)\chi \cap |p \circ \tilde{\eta}| \subset [0, 1)\chi \subset \mathrm{int}(p \circ \eta),$$

and (9.14) gives $|p \circ \tilde{\eta}| \subset \mathrm{int}(p \circ \eta)$. $\square$

CHAPTER 10

The Poincaré–Bendixson Theorem

Let $x : \mathbb{R} \to \mathbb{R}$ be a solution of equation (1.1) with phase curve $X : \mathbb{R} \ni t \mapsto x_t \in C$ in the attractor A. Recall that $\alpha(x)$ and $\omega(x_0)$ are nonempty compact connected subsets of A.

THEOREM 10.1. *Either* $\omega(x_0) = \{0\}$, *or* $0 \notin \omega(x_0)$, *and* $\omega(x_0)$ *is a periodic orbit. Either* $\alpha(x) = \{0\}$, *or* $0 \notin \alpha(x)$, *and* $\alpha(x)$ *is a periodic orbit.*

PROOF. 1. If $x(t) = 0$ for all $t \in \mathbb{R}$ then $\omega(x_0) = \{0\}$. If x is periodic and not identically zero then $0 \notin |X| = \omega(x_0)$. Assume that x is neither identically zero nor periodic, and $\{0\} \neq \omega(x_0)$. Let $\phi \in \omega(x_0) \setminus \{0\}$ be given. The solution $x(\phi)$ is slowly oscillating. All segments $x(\phi)_t, t \in \mathbb{R}$, belong to $\omega(x_0) \setminus \{0\} \subset A$. The zeroset of $x(\phi)$ has no lower bound, and all zeros of $x(\phi)$ are simple. Our first objective is to show that $x(\phi)$ is periodic. There exists a translate y of $x(\phi)$ whose zeros form a sequence $(w_j)_{-\infty}^{J(y)}$, $J(y) \in \mathbb{Z}$ or $J(y) = \infty$, with property (so), and furthermore

$$J(y) \geq 9, \quad w_0 < -1 < 0 < w_1, \quad y(t) > 0 \quad \text{for} \quad -1 \leq t \leq 0.$$

In particular,

$$y'(w_1) < 0.$$

Set $Y(t) = y_t$, for $t \in \mathbb{R}$, and $d = w_9 + 1$. Using continuous dependence on initial data we find an open neighbourhood U of y_0 such that for every $\overline{\phi} \in U \cap A$ we have $\overline{\phi}(t) > 0$ for $-1 \leq t \leq 0$, and the slowly oscillating solution $x(\overline{\phi})$ has exactly 9 consecutive zeros in the interval $(0, d)$. If z denotes the first of these zeros then

$$x(\overline{\phi})'(z) < 0.$$

Note $Y(\mathbb{R}) \subset S$ and, by Lemma 5.1, $|p \circ Y| \subset L \setminus \{0\}$. The tangent vector to the curve $p \circ Y$ at $t = 0$ is $py_0' \neq 0$ (Proposition 8.1). Choose a vector $\chi \in L \setminus \{0\}$ so that χ and py_0' are linearly independent. In particular, $py_0 + r\chi \neq 0$ for all $r \in \mathbb{R}$. Then the line

$$T = py_0 + \mathbb{R}\chi$$

is transversal to the curve $p \circ Y$ at $t = 0$, and

$$0 \notin T.$$

2. Proposition 8.2(ii) implies that there exist an open neighbourhood U_1 of y_0 in C, $\epsilon > 0$, and a continuous map $\sigma : U_1 \cap A \to (-\epsilon, \epsilon)$ such that for every $\tilde{\phi} \in U_1 \cap A$,

$$px(\tilde{\phi})_t \in T \quad \text{and} \quad |t| < \epsilon \quad \text{if and only if} \quad t = \sigma(\tilde{\phi}).$$

We have

$$\sigma(y_0) = 0.$$

Choose an open neighbourhood N of py_0 in L so that for every $\tilde{\chi} \in N \cap pA$,

$$\tilde{\chi} + a(\tilde{\chi}) \in U_1 \cap U.$$

For $\tilde{\chi} \in N \cap pA \cap T$, set $\tilde{x} = x(\tilde{\chi} + a(\tilde{\chi}))$. Using $p\tilde{x}_0 = \tilde{\chi} \in T$ we obtain

$$0 = \sigma(\tilde{\chi} + a(\tilde{\chi})) \quad \text{and} \quad p\tilde{x}_t \notin T \quad \text{for} \quad 0 < t < \epsilon.$$

3. Since $Y(0) = y_0 \in \omega(x_0)$ there is a strictly increasing unbounded sequence $(t_n)_1^\infty$ such that $X(t_n) \to Y(0)$ as $n \to \infty$. We may assume $t_{n+1} > t_n + 2\epsilon$ and $pX(t_n) \in N$ for all $n \in \mathbb{N}$. Then

$$X(t_n) = pX(t_n) + a(pX(t_n)) \in U_1 \cap U \quad \text{for all} \quad n \in \mathbb{N}.$$

For $n \in \mathbb{N}$, set $s_n = t_n + \sigma(X(t_n))$. This yields

$$pX(s_n) \in pA \cap T \quad \text{for all} \quad n \in \mathbb{N}.$$

The sequence $(s_n)_1^\infty$ is strictly increasing and tends to ∞. We have

$$\lim_{n \to \infty} X(s_n) = Y(0)$$

since

$$X(s_n) = F_A(\sigma(X(t_n)), X(t_n)), \quad X(t_n) \to Y(0),$$
$$\sigma(X(t_n)) \to \sigma(Y(0)) = 0.$$

There exists $n \in \mathbb{N}$ such that for integers $\nu \geq n$, $pX(s_\nu) \in N$, and therefore

$$pX(s_\nu) \in N \cap pA \cap T.$$

We may assume that the sequence $(|pX(s_\nu) - pY(0)|)_n^\infty$ is strictly decreasing.
4. Consider some integer $\nu \geq n$. The last statement in part 2 of the proof yields

$$pX(t) \notin T \quad \text{for} \quad s_\nu < t < s_\nu + \epsilon.$$

Let

$$s_\nu^* = \inf\{t > s_\nu : pX(t) \in T \quad \text{and} \quad \|pX(t) - pY(0)\| \leq \|pX(s_\nu) - pY(0)\|\}.$$

Then

$$s_\nu + \epsilon \leq s_\nu^* \leq s_{\nu+1}$$

since $pX(s_{\nu+1}) \in T$ and $\|pX(s_{\nu+1}) - pY(0)\| \le \|pX(s_\nu) - pY(0)\|$. By continuity,

$$pX(s_\nu^*) \in T \quad \text{and} \quad \|pX(s_\nu^*) - pY(0)\| \le \|pX(s_\nu) - pY(0)\|.$$

Define

$$\gamma(t) \;=\; pX(t) \quad \text{for} \quad s_\nu \le t < s_\nu^*,$$
$$\gamma(t) \;=\; (t - (s_\nu^* + 1))pX(s_\nu^*) + (t - s_\nu^*)pX(s_\nu) \quad \text{for} \quad s_\nu^* \le t \le s_\nu^* + 1.$$

We have $X(\mathbb{R}) \subset A \setminus \{0\} \subset S$ and $0 \notin pX(\mathbb{R})$, by Lemma 5.1. Recall $0 \notin T$. It follows that the piecewise smooth curve γ avoids $0 \in L$.

Claim: The restriction of γ to the interval $(s_\nu, s_\nu^* + 1)$ is injective. Proof. We have $x_t \ne x_s$ whenever $t \ne s$ since x is neither constant nor periodic. The fact that A is a graph over pA (Theorem 7.1(i)) yields that the curve $p \circ X$ has no self–intersections. In particular, $pX(s_\nu) \ne pX(s_\nu^*)$, and the restriction of γ to the interval $[s_\nu^*, s_\nu^* + 1]$ is injective. Recall that for $s_\nu < t < s_\nu^*$, $pX(t) \notin T$ by construction. This proves the claim.

It follows that γ is a simple closed piecewise smooth curve in $L \setminus \{0\}$. Consequently,

$$\{-1, 0, 1\} \ni \mathrm{wind}(0, \gamma) = \frac{1}{2\pi}\Re\frac{1}{i}\int_\gamma \frac{1}{z}\,dz$$

$$= \frac{1}{2\pi}\Big(\Re\frac{1}{i}\int_{I\circ p\circ X|[s_\nu, s_\nu^*]} \frac{1}{z}\,dz + \Re\frac{1}{i}\int_{I\circ\gamma|[s_\nu^*, s_\nu^* + 1]} \frac{1}{z}\,dz\Big).$$

Corollary 2.2 implies that the last integral has absolute value strictly less than π. Therefore

(10.1) $$\Re\frac{1}{i}\int_{I\circ p\circ X|[s_\nu, s_\nu^*]} \frac{1}{z}\,dz < 3\pi.$$

We have $X(s_\nu) \in U \cap A$. Therefore $X(s_\nu)(t) > 0$ for $-1 \le t \le 0$, and the zeros of x form a sequence $(z_j)_{-\infty}^J$, $J \in \mathbb{Z}$ or $J = \infty$, with property (so). We may assume that z_1 is the first zero in the interval (s_ν, ∞). Then

$$x'(z_1) < 0, \quad z_0 < s_\nu - 1, \quad x'(z_0) > 0, \quad J \ge 9, \quad z_9 < s_\nu + d.$$

(In fact, $J = \infty$. Proof: For some $t \in \mathbb{R}$, $\mathrm{sign}(y(t)) = -\mathrm{sign}(y(t - 1))$. Since $Y(t) \in \omega(x_0)$ there is a sequence $u_n \to \infty$ with $X(u_n) \to Y(t)$. For n sufficiently large, x has a zero in the interval $(u_n - 1, u_n)$.— *In the sequel we shall not make use of $J = \infty$. The reason is that we give here a proof which works also for α–limit sets, instead of ω–limit sets. The equation $J = \infty$ may be false in case we study $\alpha(x)$, under the assumptions that x is neither identically zero nor periodic, and $\{0\} \ne \alpha(x)$.*)

Claim:

(10.2) $$s_\nu^* \le s_\nu + d + 1.$$

Proof. It is sufficient to show $s_\nu^* \leq z_9 + 1$. Suppose this is false. We apply results of Chapter 9. Set $t_0 = z_0 - 1$ and consider the restriction of x to the interval $[t_0 - 1, \infty)$. Choose $\xi_0 \in \mathbb{R}$ with

$$I \circ p \circ X(z_0 + 1) = |I \circ p \circ X(z_0 + 1)| e^{i\xi_0}$$

and $\zeta_0 \in (\xi_0 - \pi, \xi_0 + \pi]$ with

$$I \circ c(0) = |I \circ c(0)| e^{i\zeta_0}.$$

Consider the angle

$$\xi(t) = \Re \frac{1}{i} \int_{I \circ p \circ X | [z_0 + 1, t]} \frac{1}{z} dz + \xi_0, \quad t \geq z_0 + 1.$$

Case 1: $J < \infty$ and $z_J + 1 \leq s_\nu^*$. Proposition 9.1 implies

$$(10.3) \qquad\qquad |\xi(s_\nu^*) - \xi(z_J + 1)| < \pi.$$

Beginning with (10.1) we get

$$
\begin{aligned}
3\pi \;>\; & \Re \frac{1}{i} \int_{I \circ p \circ X | [s_\nu, s_\nu^*]} \frac{1}{z} dz \\
=\; & \xi(s_\nu^*) - \xi(z_J + 1) + \xi(z_J + 1) - \xi(s_\nu) \quad (\text{with} \quad z_0 + 1 < s_\nu) \\
>\; & -\pi + \xi(z_J + 1) - \xi(s_\nu) \quad (\text{see } (10.3)) \\
>\; & -\pi + \beta(z_J + 1) - \pi - (\beta(s_\nu) + \pi) \quad (\text{Corollary 9.1}).
\end{aligned}
$$

Using the monotonicity of β and $s_\nu < z_2$ we infer

$$(10.4) \qquad\qquad 3\pi > 3\pi + \beta(z_J + 1) - \beta(z_2 + 1).$$

In case J is odd we use the monotonicity of β again and obtain

$$
\begin{aligned}
3\pi \;>\; & -3\pi + \beta(z_{J-1} + 1) - \beta(z_2 + 1) \\
=\; & -3\pi + \frac{J-1}{2} 2\pi - 2\pi \quad (\text{see Corollary 9.1}) \\
=\; & (J - 6)\pi,
\end{aligned}
$$

a contradiction to $J \geq 9$. In case J is even we conclude from (10.4) that

$$
\begin{aligned}
3\pi \;>\; & -3\pi + \frac{J}{2} 2\pi - 2\pi \quad (\text{Corollary 9.1}) \\
=\; & (J - 5)\pi,
\end{aligned}
$$

a contradiction.

Case 2: $J = \infty$, or $J \in \mathbb{N}$ and $s_\nu^* \leq z_J + 1$. Let j denote the largest integer such that $z_j + 1 \leq s_\nu^* \leq z_{j+1} + 1$. Then $j \geq 9$, and

$$
\begin{aligned}
3\pi &> \Re \frac{1}{i} \int_{I \circ p \circ X \mid [s_\nu, s_\nu^*]} \frac{1}{z} dz \quad \text{(see (10.1))} \\
&= \xi(s_\nu^*) - \xi(s_\nu) \quad \text{(with} \quad z_0 + 1 < s_\nu) \\
&> \beta(s_\nu^*) - \pi - (\beta(s_\nu) + \pi) \quad \text{(Corollary 9.1 is applicable since} \\
&\quad\ s_\nu^* \leq z_{j+1} + 1) \\
&> \beta(z_j + 1) - \beta(z_2 + 1) - 2\pi \quad (\beta \quad \text{is increasing;} \quad s_\nu < z_2 \\
&\quad\ \text{and} \quad z_j + 1 < s_\nu^*).
\end{aligned}
$$

As in case 1 we obtain a contradiction to $j \geq 9$.

5. Proof that y and $x(\phi)$ are periodic. The inequalities

$$
\|pX(s_\nu^*) - pY(0)\| \leq \|pX(s_\nu) - pY(0)\|, \quad \text{for} \quad \nu \geq n,
$$

imply

$$
pX(s_\nu^*) \to pY(0) \quad \text{as} \quad \nu \to \infty.
$$

Theorem 7.1(i) gives

$$
X(s_\nu^*) = pX(s_\nu^*) + a(pX(s_\nu^*)) \to pY(0) + a(pY(0)) \quad \text{as} \quad \nu \to \infty.
$$

Recall $s_\nu + \epsilon \leq s_\nu^* \leq s_\nu + d + 1$ for $\nu \geq n$ (see (10.2)). There is a subsequence $(t_{\iota(\kappa)})_1^\infty$ of $(s_\nu^* - s_\nu)_n^\infty$ which converges to some $t \in [\epsilon, d+1]$. It follows that the points

$$
F_A(t_{\iota(\kappa)}, X(s_{\iota(\kappa)})) = X(s_{\iota(\kappa)}^*), \quad \kappa \in \mathbb{N},
$$

converge to $F_A(t, Y(0)) = Y(0)$ as $\kappa \to \infty$. Therefore y is periodic.

6. Under the assumptions that x is neither periodic nor constant, and that $\{0\} \neq \omega(x_0)$, we have shown that for every $\phi \in \omega(x_0) \setminus \{0\}$ the slowly oscillating solution $x(\phi)$ is periodic. It remains to prove that $\omega(x_0)$ consists of a single periodic orbit. Fix a point $\psi \in \omega(x_0) \setminus \{0\}$. Let $\tau > 0$ denote the minimal period of $x(\psi)$, and consider the simple closed smooth curve $\eta : [0, \tau] \ni t \mapsto x(\psi)_t \in C$. By Corollary 9.2,

$$
0 \in \mathrm{int}(p \circ \eta).
$$

Suppose $|\eta| \neq \omega(x_0)$. Then there is a point $\tilde{\psi} \in \omega(x_0) \setminus |\eta| \subset A$. Either

$$
p\tilde{\psi} \in \mathrm{int}(p \circ \eta), \quad \text{or} \quad p\tilde{\psi} \in \mathrm{ext}(p \circ \eta).
$$

Consider the first case. If $p\tilde{\psi} = 0$ then the connectedness of $p\omega(x_0)$ implies that there is a point $\chi \neq 0$ in $p\omega(x_0) \setminus |p \circ \eta| \subset pA$ with $\chi \in \mathrm{int}(p \circ \eta)$. Let $\bar{\tau}$ denote the minimal period of the slowly oscillating periodic solution $x(\chi + a(\chi))$. Consider the simple closed smooth curve

$$
\bar{\eta} : [0, \bar{\tau}] \ni t \mapsto x(\chi + a(\chi))_t \in C
$$

We have $|\overline{\eta}| \cap |\eta| = \emptyset$, and $|p \circ \overline{\eta}| \subset \text{int}(p \circ \eta)$. The connectedness of $p\omega(x_0)$ implies that there is a point $\chi^* \in p\omega(x_0) \setminus (|p \circ \eta| \cup |p \circ \overline{\eta}|) \subset pA$ with $\chi^* \in \text{int}(p \circ \eta) \cap \text{ext}(p \circ \overline{\eta})$. By Corollary 9.2, $\chi^* \neq 0$. Let τ^* denote the minimal period of the slowly oscillating periodic solution $x(\chi^* + a(\chi^*))$, and consider the simple closed smooth curve

$$\eta^* : [0, \tau^*] \ni t \mapsto x(\chi^* + a(\chi^*))_t \in C.$$

The relation $\chi^* \in \text{int}(p \circ \eta)$ gives

$$|p \circ \eta^*| \subset \text{int}(p \circ \eta).$$

It follows that

$$|p \circ \eta| \subset \text{ext}(p \circ \eta^*).$$

The relation $\chi^* \in \text{ext}(p \circ \overline{\eta})$ implies

$$|p \circ \eta^*| \subset \text{ext}(p \circ \overline{\eta}).$$

Using Corollary 9.3 we get

$$|p \circ \overline{\eta}| \subset \text{int}(p \circ \eta^*).$$

Therefore there are reals $s < t$ with

$$px_s \in \text{int}(p \circ \eta^*), \quad px_t \in \text{ext}(p \circ \eta^*).$$

By continuity,

$$px_u \in |p \circ \eta^*| \quad \text{for some} \quad u \in (s, t).$$

Consequently, $x_u \in |\eta^*|$, and x is periodic, which contradicts our assumptions. The argument in case $p\tilde{\psi} \in \text{ext}(p \circ \eta)$ is analogous.

7. The proof for $\alpha(x)$ is analogous. $\square$

Next, we exclude homoclinic connections.

PROPOSITION 10.1. *Suppose x is neither identically zero nor periodic. Then $\alpha(x) \cap \omega(x_0) = \emptyset$.*

PROOF. 1. The case $\alpha(x) = \{0\}$. Suppose $\alpha(x) \cap \omega(x_0) \neq \emptyset$. By Theorem 10.1, $\omega(x_0) = \{0\}$. Hence $x_t \to 0$ as $|t| \to \infty$. There exists $s \in \mathbb{R}$ such that

$$|I(px_s)| > |I(px_t)| \quad \text{for all} \quad t < s.$$

The zeros of x form a sequence $(w_j)_{-\infty}^J, J \in \mathbb{Z}$ or $J = \infty$, with property (so). We may assume that

$$x'(w_j) > 0 \quad \text{if and only if} \quad j \in 2\mathbb{Z} \quad \text{and} \quad j \leq J.$$

Set

$$j = \max\{\iota \in 2\mathbb{Z} : \iota \leq J, w_\iota \leq s\}.$$

Since $px_t \neq 0$ for all t there exists $\epsilon > 0$ with

$$\epsilon < \min\{|I(px_t)| : w_{j-8} \leq t \leq s\}.$$

There exist $u < w_{j-8}$ with

$$|I(px_u)| = \epsilon < |I(px_t)| \quad \text{for} \quad u < t \leq s,$$

and $v > s$ with

$$|I(px_v)| = \epsilon < |I(px_t)| \quad \text{for} \quad s \leq t < v.$$

We have

$$u < w_{j-8} < w_j \leq v.$$

Set

$$k = \max\{\iota \in 2\mathbb{Z} : \iota \leq J, w_\iota + 1 \leq u\}, \quad t_0 = w_k - \frac{1}{2}.$$

The zeros of the solution $x|[t_0 - 1, \infty)$ on the interval (t_0, ∞) are given by

$$z_\iota = w_{\iota+k}, \quad \text{for} \quad \iota \in \mathbb{N}_0 \quad \text{with} \quad \iota + k \leq J.$$

We have $x'(z_0) = x'(w_k) > 0$ since k is even. Set

$$n = \max\{\iota \in \mathbb{N}_0 : z_\iota \leq v\}.$$

Obviously, $n \geq j - k$ and

$$z_0 + 1 \leq u < z_{j-k-8} < z_{j-k} \leq z_n \leq v,$$

(10.5) $\qquad |I(px_u)| = \epsilon = |I(px_v)| \quad \text{and} \quad \epsilon < |I(px_t)| \quad \text{for} \quad u < t < v.$

Consider the closed piecewise smooth curve $\gamma : [u, v+1] \to \mathbb{C}$ given by

$$\begin{aligned} \gamma(t) &= I(px_t) \quad \text{for} \quad u \leq t \leq v, \\ \gamma(t) &= \epsilon e^{i((1-t+v)\rho + (t-v)\theta)} \quad \text{for} \quad v \leq t \leq v+1 \end{aligned}$$

where

$$I(px_u) = \epsilon e^{i\theta}, \quad I(px_v) = \epsilon e^{i\rho}, \quad \theta \in [0, 2\pi) \quad \text{and} \quad \rho \in [0, 2\pi).$$

By construction, $0 \notin |\gamma|$. Theorem 7.1(i) and (10.5) imply that γ is a simple closed curve. It follows that the winding number of γ with respect to 0 is defined, and $\mathrm{wind}(0, \gamma) \in \{-1, 0, 1\}$. Hence

(10.6) $\qquad 2\pi \geq \Re\frac{1}{i}\int_\gamma \frac{1}{z}dz \geq \Re\frac{1}{i}\int_{\gamma|[u,v]} \frac{1}{z}dz - 2\pi = \xi(v) - \xi(u) - 2\pi,$

in the notation of Chapter 9.

The case $z_n + 1 \leq v$. The choice of n implies that x has no zero in the interval $(z_n, v]$. Proposition 9.1 yields

$$|\xi(v) - \xi(z_n + 1)| < \pi.$$

The last inequality and (10.6) give

$$
\begin{aligned}
5\pi \;>&\; \xi(z_n+1) - \xi(u) \\
>&\; \beta(z_n+1) - \pi - \beta(u) - \pi \quad \text{(Corollary 9.1)} \\
\geq&\; \beta(z_{j-k}+1) - \beta(z_{j-k-8}+1) - 2\pi \quad (\beta \text{ is increasing)} \\
=&\; 6\pi \quad \text{(Corollary 9.1)},
\end{aligned}
$$

which is a contradiction.

The case $v < z_n + 1$. Using (10.6) we obtain

$$
\begin{aligned}
4\pi \;>&\; \xi(v) - \xi(u) \\
>&\; \beta(v) - \pi - \beta(u) - \pi \quad \text{(Corollary 9.1)} \\
>&\; \beta(z_{j-k-2}+1) - \beta(z_{j-k-8}+1) - 2\pi \quad (\beta \text{ is increasing, and} \\
&\; z_{j-k-2}+1 \leq z_{j-k-1} < z_n \leq v) \\
=&\; 6\pi - 2\pi \quad \text{(Corollary 9.1)},
\end{aligned}
$$

which is a contradiction.

2. The case $\alpha(x) \neq \{0\}$. Theorem 10.1 implies that there is a slowly oscillating periodic solution $y : \mathbb{R} \to \mathbb{R}$, with minimal period $\tau > 0$, such that $\alpha(x) = |\eta|$ where $\eta(t) = y_t$ for $0 \leq t \leq \tau$. We may assume $y(-1) = 0 < y'(-1)$ so that, according to Corollary 9.2,

$$
(10.7) \qquad \left| \Re \frac{1}{i} \int_{I \circ p \circ \eta | [v,w]} \frac{1}{z} dz \right| < 4\pi \quad \text{for} \quad 0 \leq v < w < \tau.
$$

Suppose $\alpha(x) \cap \omega(x_0) \neq \emptyset$. Using Theorem 10.1 we infer $\omega(x_0) = \alpha(x) = |\eta|$. The zeroset of x has no upper bound since there is a sequence $t_n \to \infty$ with $x(t_n - 1) \to y(-\frac{3}{2})$ and $x(t_n) \to y(-\frac{1}{2})$ as $n \to \infty$. It follows that the zeros of x form a sequence $(w_j)_{-\infty}^{\infty}$ with property (so). We may assume $x'(w_j) > 0$ for even integers j. For all t we have $0 \neq x_t \notin |\eta|$, according to the hypothesis. Theorem 7.1(i) yields $0 \neq px_t \notin |p \circ \eta|$ for all t. Hence

$$
0 < \operatorname{dist}(|I \circ p \circ \eta|, I(px_t)) \quad \text{for} \quad w_0 + 1 \leq t \leq w_{10} + 1,
$$

and there exists $\epsilon > 0$ such that

$$
\begin{aligned}
\epsilon \;<&\; \operatorname{dist}(|I \circ p \circ \eta|, 0), \\
\epsilon \;<&\; \operatorname{dist}(|I \circ p \circ \eta|, I(px_t)) \quad \text{for} \quad w_0 + 1 \leq t \leq w_{10}.
\end{aligned}
$$

Using $x_t \to |\eta|$ as $|t| \to \infty$ we infer that there exist $s < w_0 + 1$ and $u > w_{10}$ such that

$$
\begin{aligned}
\epsilon \;=&\; \operatorname{dist}(|I \circ p \circ \eta|, I(px_s)), \\
\epsilon \;<&\; \operatorname{dist}(|I \circ p \circ \eta|, I(px_t)) \quad \text{for} \quad s < t < u, \\
\epsilon \;=&\; \operatorname{dist}(|I \circ p \circ \eta|, I(px_u)).
\end{aligned}
$$

Fix v, w in the interval $[0, \tau)$ such that

$$|I(py_v) - I(px_s)| = \epsilon = |I(py_w) - I(px_u)|.$$

2.1. Claim: If

$$\emptyset \neq (I(py_v) + [0, 1](I(px_s) - I(py_v))) \cap (I(py_w) + [0, 1](I(px_u) - I(py_w)))$$

then

$$v = w.$$

Proof. Apply Proposition 2.1 to $I(py_v), I(px_s), I(py_w), I(px_u)$. This yields $I(py_v) = I(py_w)$. Consequently, $v = w$.

2.2. We define a closed piecewise smooth curve γ in $\mathbb{C}$ as follows.

$$\gamma(t) = (s - t)I(py_v) + (t - (s - 1))I(px_s) \quad \text{for} \quad s - 1 \leq t \leq s,$$
$$\gamma(t) = I(px_t) \quad \text{for} \quad s \leq t \leq u.$$

If $v = w$,

$$\gamma(t) = (u + 1 - t)I(px_u) + (t - u)I(py_v) \quad \text{for} \quad u \leq t \leq u + 1.$$

If $v \neq w$,

$$\gamma(t) = (u + 1 - t)I(px_u) + (t - u)I(py_w) \quad \text{for} \quad u \leq t \leq u + 1,$$
$$\gamma(t) = I \circ p \circ \eta((u + 2 - t)w + (t - (u + 1))v) \quad \text{for} \quad u + 1 \leq t \leq u + 2.$$

Using the choice of ϵ, s, u, v, w and claim 2.1 we infer that in either case γ is a simple closed curve, with $0 \notin |\gamma|$. Therefore the winding number of γ with respect to $0 \in \mathbb{C}$ is defined, and $\text{wind}(0, \gamma) \in \{-1, 0, 1\}$. Hence

$$(10.8) \qquad\qquad 2\pi \geq \Re \frac{1}{i} \int_\gamma \frac{1}{z} dz.$$

Fix a zero $w_k, k \in 2\mathbb{Z}$, such that $w_k + 1 \leq s$. Set $t_0 = w_k - \frac{1}{2}$. Consider the solution $x|[t_0 - 1, \infty)$. Its zeros in the interval (t_0, ∞) are given by

$$z_j = w_{k+j}, \quad j \in \mathbb{N}_0.$$

We have

$$s < z_{-k} + 1 < \cdots < z_{-k+10} + 1 < u.$$

Corollary 9.1 yields

$$(10.9) \quad \Re \frac{1}{i} \int_{\gamma|[s,u]} \frac{1}{z} dz = \Re \frac{1}{i} \int_{I \circ p \circ X|[z_0+1,u]} \frac{1}{z} dz - \Re \frac{1}{i} \int_{I \circ p \circ X|[z_0+1,s]} \frac{1}{z} dz$$

$$\begin{aligned}
&> \beta(u) - \pi - \beta(s) - \pi \\
&\geq \beta(z_{-k+10} + 1) - \beta(z_k + 1) - 2\pi \quad (\beta \text{ is increasing}) \\
&= (10 - 2)\pi.
\end{aligned}$$

By Corollary 2.2,

$$(10.10) \qquad |\Re\frac{1}{i}\int_{\gamma|[s-1,s]} \frac{1}{z}dz| < \pi, \quad |\Re\frac{1}{i}\int_{\gamma|[u,u+1]} \frac{1}{z}dz| < \pi.$$

In case $v = w$ the inequalities (10.8), (10.9), and (10.10) lead to a contradiction.
In case $v < w$ we have

$$|\Re\frac{1}{i}\int_{\gamma|[u+1,u+2]} \frac{1}{z}dz| = |\Re\frac{1}{i}\int_{I\circ p\circ\eta|[v,w]} \frac{1}{z}dz| < 4\pi \quad \text{(see (10.7))}.$$

Using this and (10.8), (10.9), and (10.10), we arrive at a contradiction.
In case $w < v$ the same argument applies. $\square$

CHAPTER 11

Proof of Theorem 7.1(ii)

Suppose $A \neq \{0\}$. There exists $\phi \in A \setminus \{0\}$ with $\|p\phi\| \geq \|p\psi\|$ for all $\psi \in A$. We have

(11.1) $$tp\phi \notin pA \quad \text{for all} \quad t > 1.$$

Set $y = x(\phi)$.

PROPOSITION 11.1. y *is periodic.*

PROOF. Suppose y is not periodic. Then Proposition 10.1 implies $\alpha(y) \cap \omega(y_0) = \emptyset$.

Case 1: $\alpha(y) = \{0\}$. Then $0 \notin \omega(y_0)$. Theorem 10.1 guarantees that $\omega(y_0)$ is a periodic orbit, given by a simple closed smooth curve $\eta : [0, \tau] \to C$ with values in $A \setminus \{0\}$. The projected curve $p \circ \eta$ is simple closed and has values in $pA \setminus \{0\} \subset L \setminus \{0\}$, according to Theorem 7.1(i). Corollary 9.2 yields $0 \in \text{int}(p \circ \eta)$. The assumption that y is not periodic implies $py_t \notin |p \circ \eta|$ for all $t \in \mathbb{R}$ (Here we use Theorem 7.1(i) again). In particular, $p\phi \notin |p \circ \eta|$. The relation (11.1) shows that the point $p\phi$ can be connected by a curve in $L \setminus |p \circ \eta|$ to points with arbitrarily large norm. This implies $p\phi \in \text{ext}(p \circ \eta)$. On the other hand, there exist $t < 0$ so that py_t is contained in the open neighbourhood $\text{int}(p \circ \eta)$ of $0 \in p\alpha(y)$. By continuity, $py_s \in |p \circ \eta|$ for some $s \in (t, 0)$, which implies a contradiction to the assumption that y is not periodic.

Case 2: $\omega(y_0) = \{0\}$. Then $0 \notin \alpha(y)$ (Proposition 10.1), and arguments as in case 1 (with the roles of $\alpha(y)$ and $\omega(y_0)$ interchanged) lead to a contradiction.

Case 3: $\alpha(y) \neq \{0\} \neq \omega(y_0)$. Then $\omega(y_0)$ is a periodic orbit $|\eta|$ as in case 1, and $\alpha(y)$ is a periodic orbit, given by a simple closed smooth curve $\bar{\eta} : [0, \bar{\tau}] \to C$ with values in $A \setminus \{0\}$. As in case 1 we infer $p\phi \in \text{ext}(p \circ \eta)$ and $p\phi \in \text{ext}(p \circ \bar{\eta})$. Corollary 9.3 implies that either $|p \circ \bar{\eta})| \subset \text{int}(p \circ \eta)$, or $|p \circ \eta| \subset \text{int}(p \circ \bar{\eta})$. In the first case there exist $t < 0$ so that py_t belongs to the open neighbourhood $\text{int}(p \circ \eta)$ of $|p \circ \bar{\eta}| = p\alpha(y)$. By continuity, $py_s \in |p \circ \eta|$ for some $s \in (t, 0)$, which implies a contradiction to the assumption that y is not periodic. The argument in the second case is analogous. $\square$

Let $\tau > 0$ denote the minimal period of the slowly oscillating solution y, and set $\eta(t) = y_t$ for $0 \le t \le \tau$. Theorem 7.1(i) and Corollary 9.2 imply that the projected curve $p \circ \eta$ is simple closed and has values in $pA \setminus \{0\}$, and that we have $0 \in \text{int}(p \circ \eta)$.

PROPOSITION 11.2. $\text{ext }(p \circ \eta) \subset L \setminus pA$.

PROOF. Suppose there exists a point $\chi \in pA \cap \text{ext}(p \circ \eta)$. Then $\chi \ne 0$. The solution $x = x(\chi + a(\chi))$ satisfies $px_t \in \text{ext}(p \circ \eta)$ for all $t \in \mathbb{R}$. Both $p\alpha(x)$ and $p\omega(x_0)$ are contained in the set

$$\overline{\text{ext}(p \circ \eta)} \setminus (1, \infty)p\phi.$$

In particular, $0 \notin \alpha(x)$ and $0 \notin \omega(x_0)$. Theorem 10.1 implies that both $\alpha(x)$ and $\omega(x_0)$ are orbits in C of slowly oscillating periodic solutions. If x itself is periodic then

$$p\alpha(x) = p\omega(x_0) = \{px_t : t \in \mathbb{R}\} \subset \text{ext}(p \circ \eta) \setminus (1, \infty)p\phi.$$

If x is not periodic then Proposition 10.1 implies that $\alpha(x) \cap \omega(x_0) = \emptyset$. It follows that either

$$p\alpha(x) \subset \text{ext}(p \circ \eta) \setminus (1, \infty)p\phi, \quad \text{or} \quad p\omega(x_0) \subset \text{ext}(p \circ \eta) \setminus (1, \infty)p\phi.$$

In each of the cases considered there exists a slowly oscillating periodic solution $\overline{y} : \mathbb{R} \to \mathbb{R}$, with minimal period $\overline{\tau} > 0$, such that the smooth curve

$$\overline{\eta} : [0, \overline{\tau}] \ni t \mapsto \overline{y}_t \in C$$

is simple closed and satisfies

$$|p \circ \overline{\eta}| \subset \text{ext}(p \circ \eta) \setminus (1, \infty)p\phi.$$

Corollary 9.3 implies

$$|p \circ \eta| \subset \text{int}(p \circ \overline{\eta}).$$

On the other hand, the point $p\phi \in |p \circ \eta|$ can be connected by a curve in $L \setminus |p \circ \overline{\eta}|$ to points with arbitrarily large norm, which implies $p\phi \in \text{ext}(p \circ \overline{\eta})$, a contradiction. $\square$

Proposition 11.2 implies

$$pA \subset \text{int}(p \circ \eta) \cup |p \circ \eta|.$$

The next result completes the proof of Theorem 7.1(ii).

PROPOSITION 11.3. $\text{int}(p \circ \eta) \subset pA$.

PROOF. 1. We argue by contradiction and assume

$$\text{int}(p \circ \eta) \setminus pA \neq \emptyset.$$

2. Claim: There exist $\chi \in pA \setminus \{0\}$ and $\rho \in L \setminus \{0\}$ so that the solution $x = x(\chi + a(\chi))$ satisfies

$$px_0' \notin \mathbb{R}\rho \quad \text{and} \quad \chi + (0,1]\rho \subset \text{int}(p \circ \eta) \setminus pA.$$

Proof. Set $V = \text{int}(p \circ \eta) \setminus pA$. The set V is nonempty, open, and bounded. There is a point $\chi_0 \in V \setminus \{0\}$. Since V is bounded there exists $r > 1$ such that $r\chi_0 \in \partial V$. We have $r\chi_0 \neq 0$. It follows that there is a point $\lambda \in V$ with

$$\|\lambda - r\chi_0\| < \frac{1}{2}\|r\chi_0\|.$$

The relations

$$0 < \text{dist}(\lambda, L \setminus V) < \frac{1}{2}\|r\chi_0\|$$

imply that there is a point $\chi \in L \setminus V$ with

$$\|\chi - \lambda\| = \text{dist}(\lambda, L \setminus V).$$

Consequently,

$$\chi \in \partial V \subset \overline{V} \subset \overline{\text{int}(p \circ \eta)}.$$

We have $\chi \in pA$ since either

$$\chi \in |p \circ \eta| \subset pA, \quad \text{or} \quad \chi \in \text{int}(p \circ \eta) \cap (L \setminus V) \subset pA.$$

Also, $\chi \neq 0$ since

$$\|\chi\| = \|\chi - \lambda + \lambda\| \geq \|\lambda\| - \|\chi - \lambda\| > \frac{1}{2}\|r\chi_0\| - \frac{1}{2}\|r\chi_0\| = 0.$$

Set $\rho = \lambda - \chi$. For $t \in (0,1]$,

$$\|\chi + t\rho - \lambda\| = (1-t)\|\chi - \lambda\| < \|\chi - \lambda\| = \text{dist}(\lambda, L \setminus V),$$

and therefore

$$\chi + t\rho \in V.$$

The solution $x = x(\chi + a(\chi))$ is not identically zero since $\chi \neq 0$. All $px_t, t \in \mathbb{R}$, belong to pA. Proposition 8.1 yields $px_0' \neq 0$. Assume $px_0' = s\rho$ for some $s \in \mathbb{R} \setminus \{0\}$. Set

$$\epsilon = \frac{|s|}{2}\|\chi - \lambda\|.$$

There exists $\delta > 0$ such that for $|t| \leq \delta$,

$$\|px_t - px_0 - ts\rho\| \leq \epsilon|t|.$$

Consider $t \in \mathbb{R}$ with $|t| \leq \delta$ and $0 < ts < 1$. Then

$$\|px_t - \lambda\| = \|px_t - px_0 + \chi - \lambda + ts\rho - ts\rho\| \leq \|px_t - px_0 - ts\rho\| + \| - \rho + ts\rho\|$$

$$\leq \epsilon|t| + |1 - ts|\|\rho\| = (\frac{|st|}{2} + |1 - ts|)\|\chi - \lambda\| < \|\chi - \lambda\| = \text{dist}(\lambda, L \setminus V),$$

hence $px_t \in V$, a contradiction to $px_t \in pA$.

3. The case that x is periodic.   *The strategy of the proof is to produce an open band in $L \setminus pA$, along the projected periodic orbit, which separates the set pA into disjoint compact subsets so that there is a contradiction to the fact that A and pA are connected.*

Let $\overline{\tau} > 0$ denote the minimal period of x. The simple closed smooth curve $\gamma : [0, \overline{\tau}] \ni t \mapsto x_t \in C$ satisfies $|p \circ \gamma| \subset pA \setminus \{0\}$. The inclusion $\chi + (0, 1]\rho \subset \text{int}(p \circ \eta) \setminus pA$ implies

$$\chi + (0, 1]\rho \cap |p \circ \gamma| = \emptyset,$$

so that either

$$\chi + (0, 1]\rho \subset \text{int}(p \circ \gamma), \quad \text{or} \quad \chi + (0, 1]\mu \subset \text{ext}(p \circ \gamma).$$

3.1. The case $\chi + (0, 1]\rho \subset \text{ext}(p \circ \gamma)$. Then there exists $r > 0$ with

(11.2) $$\chi + (-r, 0)\rho \subset \text{int}(p \circ \gamma),$$

see Chapter 2.

3.1.1. Claim: $|p \circ \gamma| \subset \text{int}(p \circ \eta)$. Proof. Recall

$$\chi \in \overline{\text{int}(p \circ \eta)} = \text{int}(p \circ \eta) \cup |p \circ \eta|.$$

In case $\chi \in |p \circ \eta|$ we have $\chi + a(\chi) \in |\eta|$, and x is a translate of the periodic solution y. Consequently, $|\gamma| = |\eta|$, and the relation

$$\chi + (0, 1]\rho \subset \text{ext}(p \circ \gamma) \cap \text{int}(p \circ \eta)$$

implies a contradiction. Therefore, $\chi \in \text{int}(p \circ \eta)$. Using Theorem 7.1(i) we get $\chi + a(\chi) \notin |\eta|$. Hence $|\gamma| \cap |\eta| = \emptyset$. By Theorem 7.1(i), $|p \circ \gamma| \cap |p \circ \eta| = \emptyset$. Continuity and the relation $p\gamma(0) = \chi \in \text{int}(p \circ \eta)$ give $|p \circ \gamma| \subset \text{int}(p \circ \eta)$.

3.1.2. Claim: There exists $\delta > 0$ such that

$$\overline{\chi} \in \text{ext}(p \circ \gamma) \quad \text{and} \quad \text{dist}(\overline{\chi}, |p \circ \gamma|) < \delta \quad \text{imply} \quad \overline{\chi} \in L \setminus pA.$$

Proof. We argue by contradiction. Suppose the assertion is false. Then there is a sequence of points $\chi_n \in \text{ext}(p \circ \gamma) \cap pA, n \in \mathbb{N}$, with

$$\text{dist}(\chi_n, |p \circ \gamma|) < \frac{1}{n} \quad \text{for all} \quad n \in \mathbb{N}.$$

For every $n \in \mathbb{N}$ there exists $t_n \in [0, \overline{\tau}]$ such that

$$\|\chi_n - pX(t_n)\| = \text{dist}(\chi_n, |p \circ \gamma|) < \frac{1}{n}$$

(with $X(t) = x_t$, for $t \in \mathbb{R}$). A subsequence of points $t_{\nu(k)}, k \in \mathbb{N}$, converges to some point $t \in [0, \overline{\tau}]$. It follows that $pX(t_{\nu(k)}) \to pX(t)$ as $k \to \infty$, and consequently

$$\chi_{\nu(k)} \to px_t \quad \text{as} \quad k \to \infty.$$

For $k \in \mathbb{N}$, set $\psi_k = \chi_{\nu(k)} + a(\chi_{\nu(k)})$. The continuity assertion of Theorem 7.1(i) gives

$$\psi_k \to x_t \quad \text{as} \quad k \to \infty.$$

The projected curve $s \mapsto px_s$ is transversal to the line $\chi + \mathbb{R}\rho$ at $s = 0$. Proposition 8.2(ii) implies that there are an open neighbourhood U of x_t in C, $\epsilon > 0$, and a continuous map $\sigma : U \to (-t - \epsilon, -t + \epsilon)$ so that $\sigma(x_t) = -t$ and $pF_A(\sigma(\psi), \psi) \in \chi + \mathbb{R}\rho$ for all $\psi \in U \cap A$. It follows that there exists $k \in \mathbb{N}$ such that $\psi_k \in U$, and

$$(11.3) \qquad pF_A(\sigma(\psi_k), \psi_k) \in \chi + (-r, 1]\rho.$$

The relation $\chi_{\nu(k)} \in \text{ext}(p \circ \gamma) \cap pA$ implies $\psi_k \notin |\gamma|$. Therefore $F_A(s, \psi_k) \notin |\gamma|$ for all $s \in \mathbb{R}$. Consequently,

$$pF_A(s, \psi_k) \notin |p \circ \gamma| \quad \text{for all} \quad s \in \mathbb{R}.$$

Using again the relation $\text{ext}(p \circ \gamma) \ni \chi_{\nu(k)} = pF_A(0, \psi_k)$, and continuity, we obtain

$$(11.4) \qquad pF_A(s, \psi_k) \in \text{ext}(p \circ \gamma) \quad \text{for all} \quad s \in \mathbb{R}.$$

The relations (11.3), (11.2), (11.4), and $\chi \in |p \circ \gamma|$ altogether imply

$$pF_A(\sigma(\psi_k), \psi_k) \in \chi + (0, 1]\rho$$

which contradicts $\chi + (0, 1]\rho \in L \setminus pA$, see part 2.

3.1.3. Part 3.1.2 and the inclusion

$$pA \subset \overline{\text{int}(p \circ \eta)}$$

imply

$$
\begin{aligned}
pA \quad &\subset \quad \overline{\text{int}(p \circ \gamma)} \cup (pA \cap \text{ext}(p \circ \gamma)) \\
&\subset \quad \overline{\text{int}(p \circ \gamma)} \cup \{\overline{\chi} \in pA \cap \text{ext}(p \circ \gamma) : \text{dist}(\overline{\chi}, |p \circ \gamma|) \geq \delta\} \\
&\subset \quad \overline{\text{int}(p \circ \gamma)} \cup \{\overline{\chi} \in \overline{\text{ext}(p \circ \gamma)} \cap \overline{\text{int}(p \circ \eta)} : \text{dist}(\overline{\chi}, |p \circ \gamma|) \geq \delta\}.
\end{aligned}
$$

The sets $pA \cap \overline{\text{int}(p \circ \gamma)} \supset |p \circ \gamma|$ and

$$pA \cap \{\overline{\chi} \in \overline{\text{ext}(p \circ \gamma)} \cap \overline{\text{int}(p \circ \eta)} : \text{dist}(\overline{\chi}, |p \circ \gamma|) \geq \delta\}$$

are compact and disjoint. Claim 3.1.1 yields $|p \circ \eta| \subset \text{ext}(p \circ \gamma)$. From claim 3.1.2 we get

$$|p \circ \eta| \subset \{\overline{\chi} \in \overline{\text{ext}(p \circ \gamma)} \cap \overline{\text{int}(p \circ \eta)} : \text{dist}(\overline{\chi}, |p \circ \gamma|) \geq \delta\}$$

so that

$$\emptyset \neq pA \cap \{\overline{\chi} \in \overline{\text{ext}(p \circ \gamma)} \cap \overline{\text{int}(p \circ \eta)} : \text{dist}(\overline{\chi}, |p \circ \gamma|) \geq \delta\}.$$

Altogether, we arrive at a contradiction to the fact that the sets A and pA are connected.

3.2. Analogous arguments lead to a contradiction in the case $\chi + (0,1]\rho \subset$ int$(p \circ \gamma)$. One finds $\delta > 0$ with dist$(0, |p \circ \gamma|) \geq \delta$ so that pA is a subset of the union of the compact sets

$$\overline{\text{ext}(p \circ \gamma)} \cap \overline{\text{int}(p \circ \eta)},$$

$$\{\overline{\chi} \in \overline{\text{int}(p \circ \gamma)} : \text{dist}(\overline{\chi}, |p \circ \gamma|) \geq \delta\}.$$

The first of these sets contains $|p \circ \gamma|$. Corollary 9.2 implies that the second one contains 0. The sets are disjoint.

4. The case that x is not periodic. *The strategy of the previous case is not directly applicable. (In order to see this, imagine a structure in the plane which consists of the origin, a closed orbit around it, and a curve spiralling from the origin to the closed orbit. An open band along the heteroclinic curve does not split the structure into disjoint connected components.) Instead we procced as follows. The result of part 2 of the proof, namely that an open interval on a transversal to the projected phase curve $t \mapsto px_t$, with endpoint $\chi = px_0$, extends into the set $\text{int}(p \circ \eta) \setminus pA$, will be used to create on a transversal to a projected periodic orbit in pA points in the set $L \setminus pA$ which are situated between points in pA. An open set of the part of the transversal in pA is homeomorphic to an open set of the attractor $A(P)$ of the cone map P. We shall obtain a contradiction to the fact that $A(P)$ is connected.*

Proposition 10.1 yields $\alpha(x) \cap \omega(x_0) = \emptyset$. Therefore $0 \notin \alpha(x)$ or $0 \notin \omega(x_0)$.

4.1. The case $0 \notin \omega(x_0)$. Theorem 10.1 implies that $\omega(x_0)$ is the orbit in C of a slowly oscillating periodic solution $\overline{y} : \mathbb{R} \to \mathbb{R}$. Let $\overline{\tau} > 0$ denote the minimal period of $\overline{y}$, and set

$$\overline{Y}(t) = \overline{y}_t \quad \text{for all} \quad t \in \mathbb{R}, \quad \overline{\eta}(t) = \overline{y}_t \quad \text{for all} \quad t \in [0, \overline{\tau}].$$

We have $\omega(x_0) = |\overline{\eta}|$ and $p\overline{y}_0' \neq 0$. Fix some $\overline{\rho} \in L \setminus \{0\}$ with

$$p\overline{y}_0' \notin \mathbb{R}\overline{\rho}.$$

Proposition 8.2(ii) implies that there exist an open neighbourhood U of $\overline{Y}(0)$ in C, $\epsilon > 0$, and a continuous map $\sigma : U \to (-\epsilon, \epsilon)$ so that $\sigma(\overline{Y}(0)) = 0$ and

$$pF_A(\sigma(\psi), \psi) \in p\overline{Y}(0) + \mathbb{R}\overline{\rho} \quad \text{for all} \quad \psi \in U \cap A.$$

Corollary 4.1 permits to assume that for every $\psi \in U \cap A$, the tangent vector $px(\psi)_{\sigma(\psi)}' \in L$ and the vector $\overline{\rho} \in L$ are linearly independent.

Recall $X(t) = x_t$ for all $t \in \mathbb{R}$. There is a sequence of points $s_n, n \in \mathbb{N}$, with

$$s_n + 2\epsilon < s_{n+1} \quad \text{and} \quad X(s_n) \in U \cap A \quad \text{for all} \quad n \in \mathbb{N},$$

and

$$X(s_n) \to \overline{Y}(0) \quad \text{as} \quad n \to \infty.$$

For $n \in \mathbb{N}$, set $u_n = \sigma(X(s_n)) + s_n$. Then

$$u_n \to \infty \quad \text{and} \quad X(u_n) \to \overline{Y}(0) \quad \text{as} \quad n \to \infty,$$

and

$$pX(u_n) \in p\overline{Y}(0) + \mathbb{R}\overline{\rho} \quad \text{and} \quad D(p \circ X)(u_n)1 \notin \mathbb{R}\overline{\rho} \quad \text{for all} \quad n \in \mathbb{N}.$$

There exists a subsequence of points $t_k = u_{\nu(k)}, k \in \mathbb{N}$, with the following properties. For every $k \in \mathbb{N}$ there exists $r_k \in \mathbb{R}$ with

$$pX(t_k) - p\overline{Y}(0) = r_k\overline{\rho}, \quad 0 \neq \text{sign}(r_k) = \text{sign}(r_1);$$

for all $k \in \mathbb{N}$,

$$|r_{k+1}| < |r_k|;$$

$$r_k \to 0 \quad \text{as} \quad t \to \infty.$$

We may assume that all r_k are positive (Otherwise, replace $\overline{\rho}$ by $-\overline{\rho}$).

4.1.1. Claim: For every $k \in \mathbb{N}$ and every $\delta > 0$ there exists a point $\overline{\chi} \in (p\overline{Y}(0) + \mathbb{R}\overline{\rho}) \setminus pA$ with

$$\|\overline{\chi} - pX(t_k)\| < \delta.$$

Proof. Suppose the assertion is false. Then there exist $k \in \mathbb{N}$ and $\delta > 0$ such that

$$pA \supset \{\overline{\chi} \in p\overline{Y}(0) + \mathbb{R}\overline{\rho} : \|\overline{\chi} - pX(t_k)\| < \delta\}.$$

Corollary 8.2 implies that there are open neighbourhoods N of $pX(t_k)$ in L, $\overline{N}$ of χ in L, and a homeomorphism h from $N \cap pA \cap (pX(t_k) + \mathbb{R}\overline{\rho})$ onto $\overline{N} \cap pA \cap (\chi + \mathbb{R}\rho)$ with $h(pX(t_k)) = \chi$. Choose $\overline{\delta} > 0$ so that

$$pX(t_k) + (-\overline{\delta}, \overline{\delta})\overline{\rho} \subset \overline{N} \cap \{\overline{\chi} \in L : \|\overline{\chi} - pX(t_k)\| < \delta\}.$$

Then the relations

$$|\overline{r}| < \overline{\delta}, \quad \overline{h}(\overline{r}) = r, \quad \chi + r\rho = h(pX(t_k) + \overline{r}\overline{\rho})$$

define a continuous injective map $\overline{h} : (-\overline{\delta}, \overline{\delta}) \to \mathbb{R}$ with $\overline{h}(0) = 0$, and the image $\overline{h}((-\overline{\delta}, \overline{\delta}))$ is an open interval containing $0 \in \mathbb{R}$. This contradicts the fact that

$$\chi + (0, 1]\rho \subset L \setminus pA,$$

see part 2 of the proof.

4.1.2. For some $t \in \mathbb{R}$, we have $\overline{Y}(t) \in K$. Corollary 8.1 implies that there exist open neighbourhoods N of $p\overline{Y}(0)$ in L and U of $\overline{Y}(t)$ in C, and a homeomorphism h from $N \cap pA \cap (p\overline{Y}(0) + \mathbb{R}\overline{\rho})$ onto $U \cap A(P)$. Using part 4.1.1 we infer that there exist $\delta > 0$ with

$$p\overline{Y}(0) + (-\delta, \delta)\overline{\rho} \subset N,$$

an integer $k \in \mathbb{N}$, and reals r, r^* so that

$$0 < r < r_k < r^* < \delta$$

and

$$p\overline{Y}(0) + r\overline{\rho} \in L \setminus pA, \quad p\overline{Y}(0) + r^*\overline{\rho} \in L \setminus pA.$$

The compact set

$$M = pA \cap (p\overline{Y}(0) + [r, r^*]\overline{\rho})$$

in the topological space

$$pA \cap (p\overline{Y}(0) + (-\delta, \delta)\overline{\rho})$$

is identical with the set

$$pA \cap (p\overline{Y}(0) + (r, r^*)\overline{\rho}).$$

It follows that M is also an open subset of the space $pA \cap (p\overline{Y}(0) + (-\delta, \delta)\overline{\rho})$. We have $pX(t_k) \in M$ since

$$pX(t_k) \in pA \quad \text{and} \quad pX(t_k) = p\overline{Y}(0) + r_k\overline{\rho}.$$

Also,

$$p\overline{Y}(0) \in (pA \cap (p\overline{Y}(0) + (-\delta, \delta)\overline{\rho})) \setminus M.$$

The set $h(M) \subset A(P)$ is compact since the map

$$pA \cap (p\overline{Y}(0) + (-\delta, \delta)\overline{\rho}) \ni \overline{\chi} \mapsto h(\overline{\chi}) \in C$$

is continuous. The set $h(M)$ is an open subset of the topological space $A(P)$ since the homeomorphism h maps the open subset M of the space $N \cap pA \cap (p\overline{Y}(0) + \mathbb{R}\overline{\rho})$ onto an open subset of $U \cap A(P)$, and therefore onto an open subset of $A(P)$. It follows that

$$A(P) = h(M) \cup (A(P) \setminus h(M))$$

is a decomposition into disjoint open subsets. Both sets are nonempty since $h(pX(t_k)) \in h(M)$, $h(p\overline{Y}(0)) \in A(P) \setminus h(M)$. This contradicts the fact that $A(P)$ is connected.

4.2. The proof in the case $0 \notin \alpha(x)$ is analogous. $\quad\square$

References

1. H. Cartan, *Théorie élémentaire des fonctions analytiques d'une ou plusieurs variables complexes*, Hermann, Paris, 1961.
2. O. Diekmann, S. van Gils, S. Verduyn Lunel and H.O. Walther, *Delay Equations: Complex, Functional and Nonlinear Analysis*, (in preparation).
3. J. Dieudonné, *Foundations of Modern Analysis*, Academic Press, New York, 1960.
4. C.C. Fenske, personal communication, 1978.
5. J.K. Hale, *Theory of Functional Differential Equations*, Springer, New York, 1977.
6. J.K. Hale, *Asymptotic Behaviour of Dissipative Systems*, Amer. Math. Soc., Providence, R.I., 1988.
7. J.K. Hale, X.B. Lin, *Symbolic dynamics and nonlinear semiflows*, Ann. Mat. Pura Appl. (4) **144** (1986), 229–259.
8. A.F. Ivanov, B. Lani–Wayda, H.O. Walther, *Unstable hyperbolic periodic solutions of differential delay equations*, Recent Trends in Differential Equations (R.P. Agarwal ed.), World Scientific, Singapore, 1992, pp. 301–316.
9. J.L. Kaplan, J.A. Yorke, *On the stability of a periodic solution of a differential delay equation*, SIAM J. Math. Anal. **6** (1975), 268–282.
10. B. Lani–Wayda, H.O. Walther, *Chaos for delayed negative feedback*, preprint, 1993.
11. J. Mallet-Paret, *Morse decompositions for differential delay equations*, J. Differential Equations **72** (1988), 270–315.
12. J. Mallet-Paret, R.D. Nussbaum, *Boundary layer phenomena for differential–delay equations with state–dependent time lag I*, Arch. Rat. Mech. Analysis **120** (1992), 99–146.
13. J. Mallet-Paret, G. Sell, (in preparation).
14. J. Mallet-Paret, H.L. Smith, *The Poincaré-Bendixson theorem for monotone cyclic feedback systems*, J. Dynamics Diff. Equations **2** (1990), 367–421.
15. A. Neugebauer, *Invariante Mannigfaltigkeiten und Neigungslemmata für Abbildungen in Banachräumen*, Diploma thesis, Universität München, 1988.
16. R.A. Smith, *Poincaré-Bendixson theory for certain retarded functional differential equations*, Differential and Integral Equations **5** (1992), 213–240.

17. H.O. Walther, *On instability, ω–limit sets and periodic solutions of nonlinear autonomous differential delay equations*, Lecture Notes in Math., vol. 730, Springer–Verlag, Berlin and New York, 1979, pp. 489–503.

18. H.O. Walther, *Density of slowly oscillating solutions of $\dot{x}(t) = -f(x(t-1))$*, J. Math. Anal. Appl. **79** (1981), 127–140.

19. H.O. Walther, *An invariant manifold of slowly oscillating solutions for $\dot{x}(t) = -\mu x(t) + f(x(t-1))$*, J. Reine Angew. Math. **414** (1991), 67–112.

20. H.O. Walther, *Unstable manifolds of periodic orbits of a differential delay equation*, Oscillation and Dynamics in Delay Equations (J.R. Graef and J.K. Hale, eds), Amer. Math. Soc., Providence, R.I., 1992, pp. 177–240.

21. H.O. Walther, *A differential delay equation with a planar attractor*, Proc. Internat. Conf. on Differential Equations (Marrakech 1991), (to appear).

MATHEMATISCHES INSTITUT,UNIVERSITÄT MÜNCHEN,D 80333 MÜNCHEN,GERMANY
E-mail address: walther@rz.mathematik.uni-muenchen.de

Editorial Information

To be published in the *Memoirs*, a paper must be correct, new, nontrivial, and significant. Further, it must be well written and of interest to a substantial number of mathematicians. Piecemeal results, such as an inconclusive step toward an unproved major theorem or a minor variation on a known result, are in general not acceptable for publication. *Transactions* Editors shall solicit and encourage publication of worthy papers. Papers appearing in *Memoirs* are generally longer than those appearing in *Transactions* with which it shares an editorial committee.

As of October 3, 1994, the backlog for this journal was approximately 5 volumes. This estimate is the result of dividing the number of manuscripts for this journal in the Providence office that have not yet gone to the printer on the above date by the average number of monographs per volume over the previous twelve months, reduced by the number of issues published in four months (the time necessary for preparing an issue for the printer). (There are 6 volumes per year, each containing at least 4 numbers.)

A Copyright Transfer Agreement is required before a paper will be published in this journal. By submitting a paper to this journal, authors certify that the manuscript has not been submitted to nor is it under consideration for publication by another journal, conference proceedings, or similar publication.

Information for Authors and Editors

Memoirs are printed by photo-offset from camera copy fully prepared by the author. This means that the finished book will look exactly like the copy submitted.

The paper must contain a *descriptive title* and an *abstract* that summarizes the article in language suitable for workers in the general field (algebra, analysis, etc.). The *descriptive title* should be short, but informative; useless or vague phrases such as "some remarks about" or "concerning" should be avoided. The *abstract* should be at least one complete sentence, and at most 300 words. Included with the footnotes to the paper, there should be the 1991 *Mathematics Subject Classification* representing the primary and secondary subjects of the article. This may be followed by a list of *key words and phrases* describing the subject matter of the article and taken from it. A list of the numbers may be found in the annual index of *Mathematical Reviews*, published with the December issue starting in 1990, as well as from the electronic service e-MATH [**telnet e-MATH.ams.org** (or **telnet 130.44.1.100**). Login and password are **e-math**]. For journal abbreviations used in bibliographies, see the list of serials in the latest *Mathematical Reviews* annual index. When the manuscript is submitted, authors should supply the editor with electronic addresses if available. These will be printed after the postal address at the end of each article.

Electronically prepared manuscripts. The AMS encourages submission of electronically prepared manuscripts in $\mathcal{A}_{\mathcal{M}}\mathcal{S}$-TeX or $\mathcal{A}_{\mathcal{M}}\mathcal{S}$-LaTeX because properly prepared electronic manuscripts save the author proofreading time and move more quickly through the production process. To this end, the Society has prepared "preprint" style files, specifically the amsppt style of $\mathcal{A}_{\mathcal{M}}\mathcal{S}$-TeX and the amsart style of $\mathcal{A}_{\mathcal{M}}\mathcal{S}$-LaTeX, which will simplify the work of authors and of the production staff. Those authors who make use of these style files from the beginning of the writing process will further reduce their own effort. Electronically submitted manuscripts prepared in plain TeX or LaTeX do not mesh properly with the AMS production systems and cannot, therefore, realize the same kind of expedited processing. Users of plain TeX should have little difficulty learning $\mathcal{A}_{\mathcal{M}}\mathcal{S}$-TeX, and LaTeX users will find that $\mathcal{A}_{\mathcal{M}}\mathcal{S}$-LaTeX is the same as LaTeX with additional commands to simplify the typesetting of mathematics.

Guidelines for Preparing Electronic Manuscripts provides additional assistance and is available for use with either $\mathcal{A}_{\mathcal{M}}\mathcal{S}$-TeX or $\mathcal{A}_{\mathcal{M}}\mathcal{S}$-LaTeX. Authors with FTP access may obtain *Guidelines* from the Society's Internet node e-MATH.ams.org (130.44.1.100). For those without FTP access *Guidelines* can be obtained free of charge from the e-mail address guide-elec@ math.ams.org (Internet) or from the Customer Services Department, American Mathematical Society, P.O. Box 6248, Providence, RI 02940-6248. When requesting *Guidelines*, please specify which version you want.

At the time of submission, authors should indicate if the paper has been prepared using $\mathcal{A}_{\mathcal{M}}\mathcal{S}$-TeX or $\mathcal{A}_{\mathcal{M}}\mathcal{S}$-LaTeX. The *Manual for Authors of Mathematical Papers* should be consulted for symbols and style conventions. The *Manual* may be obtained free of charge from the e-mail address cust-serv@math.ams.org or from the Customer Services Department, American Mathematical Society, P.O. Box 6248, Providence, RI 02940-6248. The Providence office should be supplied with a manuscript that corresponds to the electronic file being submitted.

Electronic manuscripts should be sent to the Providence office immediately after the paper has been accepted for publication. They can be sent via e-mail to pub-submit@math.ams.org (Internet) or on diskettes to the Publications Department, American Mathematical Society, P.O. Box 6248, Providence, RI 02940-6248. When submitting electronic manuscripts please be sure to include a message indicating in which publication the paper has been accepted.

Two copies of the paper should be sent directly to the appropriate Editor and the author should keep one copy. The *Guide for Authors of Memoirs* gives detailed information on preparing papers for *Memoirs* and may be obtained free of charge from the Editorial Department, American Mathematical Society, P.O. Box 6248, Providence, RI 02940-6248. For papers not prepared electronically, model paper may also be obtained free of charge from the Editorial Department.

Any inquiries concerning a paper that has been accepted for publication should be sent directly to the Editorial Department, American Mathematical Society, P.O. Box 6248, Providence, RI 02940-6248.

Recent Titles in This Series

(Continued from the front of this publication)

(See the AMS catalog for earlier titles)

Recent Titles in This Series

(Continued in the back of this publication)